Education, Human Capital and Research Capacity for African Integrated Development

Education, Human Capital and Research Capacity for African Integrated Development

Editors

Mammo Muchie & Angathevar Baskaran

AFRICA WORLD PRESS
Trenton | London | Cape Town | Nairobi | Addis Ababa | Asmara | Ibadan | New Delhi

AFRICA WORLD PRESS
541 West Ingham Avenue | Suite B
Trenton, New Jersey 08638

Book design: Dawid Kahts
Cover art: Ananthy Baskaran

Library of Congress Cataloging-in-Publication Data

Names: Muchie, Mammo, editor, author. | Baskaran, Angathevar, editor, author.
Title: Education, human capital and research capacity for African integrated development / editors, Mammo Muchie & Angathevar Baskaran.
Description: Trenton : Africa World Press, 2017. | Includes bibliographical references and indexes.
Identifiers: LCCN 2017007918| ISBN 9781569025635 (hb : alk. paper) | ISBN 9781569025642 (pb : alk. paper)
Subjects: LCSH: Economic development--Africa, Sub-Saharan. | Technological innovations--Africa, Sub-Saharan. | Education, Higher--Africa, Sub-Saharan. | Research--Africa, Sub-Saharan.
Classification: LCC HC800.Z9 .T4334 2017 | DDC 338.0640967--dc23
LC record available at https://lccn.loc.gov/2017007918

This book is dedicated to

Kegnazmatch Asmetche Zenbe Engida and Yalew Getahun who fought the fascist war and valiantly left this world to save Ethiopia!
&
All freedom fighters across Africa

Contents

Acknowledgements

We would like to thank the DST/NRF SARChI (Innovation and Development) at the Tshwane University of Technology, Pretoria, South Africa for initiating the Science, Technology, Innovation Studies series of books in collaboration with African World Press.

We recognize and appreciate the support given by Dr. Jideofor Adibe, Adonis & Abbey Publishers Ltd, London, United Kingdom, which enabled publication of this book.

We also would like to thank the African Journal of Science, Technology, Innovation and Development (AJSTID) and extend our fullest appreciation and thanks to all the contributing authors.

Our sincere thanks to Ananthy Baskaran, the cover designer, and Jan B. Mwesigwa (www.janmwesigwa.com) our initial layout artist for their hard work and dedication to the book project.

Finally, we express our deepest appreciation and gratitude to Mr. Kassahun Checole and Mr. Girma A. Demeke, Africa World Press, Inc. & The Red Sea Press, Inc. for their cooperation and support during the publication process. We also thank all the publications staff at Africa World Press, Inc. & The Red Sea Press, Inc.

The DST/NRF Research Chair at Tshwane University of Technology and Africa World Press will promote the science, technology, innovation studies book series applying the principle to promote innovation and knowledge on and for Africa.

PREFACE:

Education and Research for Human Capital

Mammo Muchie & Angethvar Baskaran

There are two critical capitals that are essential to manage systematically, efficiently and effectively all other capitals. These are human capital and trust capital. Deficiences in these two capitals bring problems to social, economic and natural sustainable development. Both humanity and nature will be at risk when social and human capitals fail to be effective in managing social relations, economic development and nature protection. Other capitals are essential such as financial capital, organisational capital, natural capital and physical capital but they all need necessary and sufficient social and human capital to contribute to social and economic transformation.

Human capital is built through quality education. Knowledge through education is critical to build the human capital resources to foster development. Education is not just an option to choose or not to choose. All human beings without any exception and whoever they are should be educated. Life without education is empty. Everyone should make life an education journey without any reservation.

When education plays the role like life itself, the critical capital that is needed human capital will grow and contribute to promote all the natural, organisational, financial and physical capitals. If human capital is the necessary driving force for other capitals, education which is the foundation of human capital becomes the critical power to promote all the capitals.

Education requires both training and research capacity to deepen with imaginative resources to develop the knowledge sphere to become sharper, broader and more inclusive with new ideas, methods and approaches. Research enhances education and education builds human capital and human capital facilitates all other capitals.

For Africa we need to arrange systematically research capacity, education, human capital and social, financial, natural, physical and organisation capitals. We need to know how to integrate them by recognising both necessary and sufficient conditions.

It is reported that there are 130 million young people worldwide not being enrolled in school. Imagine how many potential nobel laurates are not in school and it is not just their loss, it is also the loss of humanity not to have the opportunity to broaden and deepen the knowledge frontier. There is no doubt most of these young people are from the developing world. There is also problems with the quality of education. In much of Africa the curriclum does not refelect the African context and a lot of the resources are borrowed and both the graduates and the education may not be useful to generate quality human capital. The colloboration amongst the key actors and stakeholders appears to be very shallow. The epistemological foundation of the education system comes from the colonial times and no serious effort has been made to re-think, re-design and re-found education on a strong-African context that is critically important to address. It is not time spent in learning that matters but the quality of research and education that learners experience and achieve that goes a long way to make a difference in building human capital.

This book is very important for it links education, research capacity and human capital for the promotion of the African integrated development. A unified and hybrid imagination is critically important to create an innovative integrated development trajectory for Africa. Quality and not quantity in education needs to be selected to promote Africa's integrated development path. The contributions highlight the critical importance of human capital for sustainable and integrated African development.

CHAPTER 1

Challenges Towards Building Education, Human Capital and Research Capacity for Integrated African Development: Introduction

Mammo Muchie & Angathevar Baskaran

An emergent field of science, technology, innovation for development (STI4D) or what originated in 2003 the Global Research Network on Innovation, development and comptence building (GLOBELICs) simply renders as Innovation and Development (I&D) has opened up opportunities to incorporate explicitly research, knowledge, invention, creativity and innovation in changing the economic condition of underdevelopment that many economies face in the world in general and in Africa in particular. The explicit appropriation and incorporation of knowledge, invention and innovation to revitalise and dynamise the development debate by adding new and fresh perspectives and directions is indeed timely and necessary. The creative combination of innovation and development will generate new insights and approaches in examining the unacceptable continuation of the state of underdevelopment that has been plaguing many low income economies in Africa and elsewhere.

One important constraint is that innovation theory has evolved without considering explicitly the development context and issues. It evolved mainly as part of evolutionary economic theory, which in turn came in opposition to neo classical economic theory. The theoretical effort to link innovation theory with development economics may require that both innovation theory and theory of development economics are re-thought and reconfigured. Something like the development economics of innovation may need to be developed.

What remains a puzzle is that combining innovation systems with development economics remains problematic and finding ways to make the synthesis that generates insights, learning, theorems and workable policy directions remain a big challenge. One way to make the combination work is to make the separate communities from innovation and development economics to join in GLOBELICS or similar networks at regional level such as ASIALICS and AfricaLICs. Whilst the move to make the combination is inspiring, the actual theoretical synthesis and its impact remain to be realised. Disciplinary hybrid imagination is needed.

In the African context, the combination of innovation with development is a big challenge. It is not easy to do it. The main reason is that Africa's economic development has been constrained by the load it still suffers from the colonial legacy. The states as they exist are divided and when they form regions, they still remain divided. There are claims that Africa's fortunes are changing and conflicts are decreasing. There is now the Africa rise narrative. There are some economies that are growing above 7%. Democratisation is slowly and surely growing. Technologically mobile phone users are growing and the ICT revolution in Africa may help Africa to leapfrog and join as a player rather than a victim of the division of labour in the world economy. Whilst these optimistic scenarios are useful and may counter the pessimism of diagnosis and pessimism of prescription that Africa has faced for centuries, we need to appreciate that the key to Africa's development remain the ability for Africa to create an African united economy. The existing states or the regions they suffer from a formidable deficit, that is, failure to 'Put Africa

First' and failure to formulate a united policy and strategy to deal with the colonial legacy and the new and current challenges Africa is facing today. Research on African innovation and development is needed at various levels to demonstrate what would work most effectively in the specific African context.

The African Journal of Science, Technology, Innovation and Development (AJSTID), started by us in 2009, is one of the journals in Africa that is inspired by the effort to create systematic research on innovation, development, learning and competence building systems in developing economies in general and the African economy in particular. It welcomes scientific research that twins the theory of the development economics of innovation with theories of evolutionary economics and science, technology and innovation systems. It recognises the enormous research benefits that twinning established research scientists from the North with the South brings. It welcomes co-publication of research outputs that twins work by both top researchers in the North and those in the South, on the interdisciplinary fields of innovation studies with development studies. AJSTID encourages especially research publications from emergent researchers with senior researchers that enrich the multidisciplinary field of innovation and development. The combination of research excellence coming from the global pool of researchers and original and quality research from African emergent and senior researchers has produced opportunities for creating measurable research capacity development in Africa. Excellence in research is encouraged by trying to incentivise top scholars in the North to spread and diffuse research interactive skills with the emergent researchers in the South. It is through the explicit realisation when such top level quality and original research circulates without barriers and active networking, the opportunities for multiplying research capacity at the bottom of the research pyramid in the developing world and especially Africa grows.

What are some of the challenges that African countries face in establishing the scientific enterprise and in moving from science to the elusive goal of growth and development? The book explores some of the efforts, challenges in and outcomes from attempting to use science, technology, and innovation (STI) for socio-economic development in African countries, particualry the challenges faced by them towards building education, human capital and research capacity for integrated development. This book features research contributions that focused primarily on this issue.

The contributing authors approach their discussions through different disciplinary and theoretical lenses, but using the central concept of system of innovation for building education, human capital and research capacity for integrated African development. The concept was developed to overcome weaknesses perceived in neoclassical explanations to connect outputs of scientific activity with economic growth. It provides a useful framework to examine the linkages of research (knowledge creation) to its eventual contribution to output growth. Standard economic approaches fail to take into account a number of issues such as feedbacks between producers and users in the innovation process; and adherence to the assumptions of profit maximization. While it is still not yet a fully developed theory, the approach helps to identify the components and processes involved in innovation and the potential to contribute to growth. The systems of innovation approach focuses on the range of entities / institutions / activities involved in R&D and commercialization of the output. However, the approach is not focused solely on the entities involved in the innovation processes but it is also concerned about the relationships and interactions between the components in the system.

Knowledge is a central element of the system of innovation approach; and as a consequence, entities that produce knowledge (universities, research institutes, and industry) are an important feature of the system. However, the systems of innovation approach do not limit consideration solely to knowledge produced though research or activities associated with science based innovations but can also accommodate knowledge produced by other sources. The process of learning, "learning by doing", using and interacting are also taken into consideration in the analytical reasoning. The innovation process rests not just on knowledge produced through research but it also depends on innovations taking place in processes, organizations, or markets.

This book presents research work across Africa on the following topics:

- Relevance of technical and vocational education in national development strategy
- Efforts to build sustainable African research and innovation capacity
- A nanostep towards building a sustainable interdisciplinary research and educational program
- Challenges and opportunities of re-inventing higher education as centres of research capacity building
- Undergraduates' access to E-learning: policy implications
- Demystifying and demythologizing technology and technology education
- Local universities as engines for innovation, take-off & expansion of regional development in southern economies
- Constructing multidimensional education and health welfare indexes
- Investigation into the factors affecting corporate sustainability of public further education and training colleges
- Human capital development, technological capabilities and national development

Victor Dike argues, based on the experience of Nigeria, that without addressing the challenges facing education in general, and technical and vocational education in particular, the youth will continue to face soaring unemployment, underemployment and poverty and the nation will remain underdeveloped. He highlights the failure of government to properly fund education and to restructure the curriculum to prepare the youth for the world of work as well as to create employment for the millions of graduates churn out by the educational institutions yearly.

Joseph M. Kizza, Mammo Muchie & Timothy Waema decry the poor performance of African countries, with exception of South Africa and Egypt, compared to other countries in other parts of the globe, with respect to research and innovation generation. They describe a number of initiatives the authors have been involved in to create a culture and build sustainable African research and innovation capacity essential for the development of the continent. The chapter ends by outlining how these initiatives could benefit the SIGCSE community and proposes three strategies to enhance Africa's research capacity: establish academic research networking and training; identify cognate fields from social and technical sciences with research programmes and research themes to create, acquire, build and use a body of knowledge for both the enhancement of research and quality training; and organisation of a number of books, journals and other publication that will be channels of scholars research ideas and also to provide immediate outlet in Africa. Finally, they also highlight the likely impacts of these initiatives on African ICT and STI research capacity.

Myron N.V. Williams, Paul M. Brown, Ishrat M. Khan, Alfred Msezane, James Reed, Laurisa London & Malik Maaza discuss the experience of a pilot project International Research Experience for Undergraduates (IREU) conducted jointly by Clark Atlanta University's (CAU) Center for Functional Nanoscale Materials (CFNM) and iThemba LABS and the African Laser Centre in South Africa. The students involved in the project did laboratory and theoretical research to create gold/polymer composite nanostructures that may function in nanobiosensors. The project found that the students involved in the project had an experience both at CAU and in South Africa, where they met and collaborated with scientists and students from various South African Institutions and from all over Africa.

Mammo Muchie & Angathevar Baskaran discuss how research capacity that can promote a sustained research environment and research identity as the defining attributes of universities can be enhanced based on a case study of a research network on the economics of innovation, learning and competence building that we have come to be familiar with. They argue that the research systems do not exist in a vacuum and they are connected to a broader system of innovation. The more robust a given national system of innovation, the more likely it is to impact positively on research system development. Conversely, the weaker a given system of innovation, the more unlikely the broader context would be conducive to research system development and evolution. The case study of the Global Network for Economics of Learning, Innovation, and Competence Building Systems (GLOBELICS) experiment is used, in the limited sense, to illustrate and exemplify a global research networking that is inclusive of the South by moving those from the North to go to the South rather than the other way around. The work aims to contribute insight into how research capacity through networks that complement the university's formal training based research work can work.

Nwaocha Vivian Ogochukwu explores students' access to e-learning policy implications for tertiary institution administration in Nigeria. She used data gathered from randomly chosen 500 undergraduates from five tertiary institutions. Data obtained were analysed using descriptive statistics and population t-test of single means. Results of the analysis signified that the number of students in Nigerian tertiary institutions who have access to e-learning technology was insignificant. It is expected that the results obtained from this study will assist policy makers and University administrators find solutions to making e-learning technology more accessible to undergraduates at all levels.

Thabo I Pudi, Vuyisile Msila, Cheryl le Roux, Agnieta B Pretorius, Coneth Richards & Marius Dieperink argue that there is widespread misunderstanding and misrepresentation of what Technology and Technology Education (TE) are. They look at some of these misunderstandings and misinterpretations that teachers hold about Technology and TE and undertake to explain their rationale. They perceive that broadening knowledge by delving into the different conceptions of what Technology and consequently TE is, is tantamount to their demystification and demythologization. To investigate TE teachers' understanding of Technology Education and its implementation, they divided the study into three broad themes, that is, the concepts *Technology*, *TE* and *TE in the curriculum*, and into sub-themes. Based on these themes and sub-themes, a questionnaire survey was constructed to collect both qualitative and quantitative data. This study revealed lack of teachers' understanding of the concepts Technology and TE and as a result lack of proper understanding of how TE should be implemented within the curriculum. The existence of myths and stigmas was also found.

Ahmed Driouchi & Nada Zouag aim to show that universities can be engines for local development and tools for innovation take-off. Previous contributions to literatures on this subject have already shown the positive effects of regional sources of knowledge on local development. Using data on developed, developing and emerging countries, regression analysis they show that developing economies do have room for innovation and local development, as this can be further provided by regional universities and schools. These potential gains have been expressed to be higher for developing and emerging countries. These results imply that developing and emerging countries can enhance their local and overall development and can ensure improved innovation possibilities through the promotion of local universities and schools but these sources of skills and knowledge need to be tied with the local needs of the population.

Boniface Ngah Epo & Francis Menjo Baye review the concept of Multiple Correspondence Analyses (MCA) in the area of multidimensional poverty measurement and apply this analysis to multidimensional welfare in Cameroon. Use is made of the second (2001) and third (2007) Cameroon Consumption Household Survey. Results indicated that modalities like knowing how to read and write, sector of consultation, type of sanitary centre, appreciation of health status, cost of going to the hospital, cost of education, time and distance to the nearest school or hospital affect capabilities like education and health and therefore household economic well-being. Policy suggestion is that defining adequate policy mixes that blend human capital investment, in education and health, and complementary investments like infrastructural development that go together with human investment should be suggested to policy makers as ways of consolidating development and increasing household welfare.

What factors lead to the implementation and success of corporate sustainability programs and policies in public Further Education and Training Colleges (FET)? To find out Manager M. Muswaba & Zeleke Worku examine FET institutions' self-reported corporate sustainability initiatives in Gauteng Province of South Africa. Related literature showed that the quality of services on corporate sustainability provided by the college council is influenced by stakeholders, social justice, economic and corporate cultural factors. Findings revealed that involvement of stakeholders in sustainability planning processes created greater willingness to implement sustainability projects. Results also revealed that FET Council Boards should be effective change agents and must have the ability to visualise, verbalise, and set into motion clear, focused action plans related to corporate sustainability.

Victor E. Dike argues that the neglect of education, especially technical and vocational education and science-based technology education in Nigeria has made a review of the poor investment in human capital development and national development rather compelling. He highlights that the political leaders have always raised the people's hope by promising that they would give priority attention to human capital development and national development and empower the citizens particularly the poorly educated, unskilled, and unemployed youths with relevant human skills capital (HSK) and entrepreneurial skills to enable them to contribute profoundly to national development. Yet the political leaders have over the years failed to adequately fund education and strengthen the infrastructure and institutions that would drive the economy and create employment for the teeming population. He argues that the reality though is that Nigeria cannot become an industrialized society without investing handsomely on human capital development (education and health) and technological capabilities.

The contributions together highlight the significance of education in building human capital for economic development. It matters very much how to manage the accumulation of human capital in relation to all other capitals in order to promote sustainable and integrated economic development. Human capital with skills and knowledge is critical for African social, economic and sustainable development in this time of the digital and knowledge economy. Research, education, human capital together can be dynamic agents to make people acquire knowledge to encourage a path of an integrated African development to become a reality through the promotion of interactive learning, creativity and innovation by promoting the wellbeing of both people and nature.

CHAPTER 2

The Relevance of Technical and Vocational Education in Nigeria's Development Strategy:

A New Paradigm for Effective Economic Transformation

Victor E. Dike

1. Introduction

Technical and vocational education (TVE) has been an integral part of national development strategies in many societies because of its impact on productivity and economic development (Wirth, 1972; McNabb, 1997). It is concerned with the acquisition of skills and knowledge for employment and sustainable livelihood (Maclean and Wilson, 2009). Secondary schools in industrialized nations have vocational and technical education departments that prepare the students for the world of work. Since huge contributions are made by the skilled workers toward organizational growth and development, the organizations in development-conscious nations spend billions of dollars every year on-programs aimed at upgrading of skills and retraining of their employees. The present-day 'knowledge-driven' global economy requires an educational program that provides not only academic knowledge and job skills but problem-solving skills, creative and critical minds, and socially responsible citizens.

Despite the contributions of vocational and technical education to national development the leaders of Nigeria have not given due attention to the TVE sector, resulting in the neglect of vocational and technical educational institutions and the youths shying away from technical and vocational studies. Because of the neglect of this important sector, most of the secondary school graduates (and some of those who managed to pass through the university) lack the skills and knowledge to compete effectively in the rather tight labor market and thus loiter around in the villages and cities from dawn to dusk looking for jobs that are not available.

Vocational education is a practical (non-academic) instruction that gives the learners specific occupational skills for a trade or vocation and for entry-level jobs. It involves mostly hands-on laboratory process and on-the-job training for proficiency in manual skills. Vocational education takes the needs of the society and individual into consideration. However, technical education prepares students for acquisition of skills as well as basic scientific knowledge. One of the primary objectives of technical education is providing employability and workplace skills and the preparation of graduates for occupations that are above skilled crafts but below the scientific or engineering professions. It also prepares graduates for continuing education and often involves the mastery of a body of knowledge and skills that can be applied in a practical way (Doll and Peto, 1985; Asche, 1991; Warnat, 1991; Gleeson & Mardle, 1980; Washington-Office of Superintendent of Public Instruction (no date); UNEVOC, 1996:140; and Maclean and Wilson, 2009).

Specifically, vocational and technical education gives individuals the skills to learn and become productive citizens and for advancement in the workplace (Oni, 2006). Some scholars perceive technical and vocational education as one of the 'bulwarks of social efficiency' as the preparation of a well-trained workforce is a *sine qua non* of 'an efficient society' (Camp, 1982:11-18; 1983:10-19). Thus because of the unrelenting changes in the new global economy one may not be relevant in the labor market in future without certain level of technical skills. The neglect of vocational and technical education in Nigeria leads to the shortage of skilled technicians to maintain the nation's critical infrastructure and to tackle its developmental challenges.

This paper seeks to identify the relevance of vocational and technical education and training on socio-economic and political development and the extent to which it can effectively prepare the youth for future positions in business and industry. Specifically, it focuses on the dearth of skilled technical manpower in Nigeria and argues that vocational and technical education holds the key to national development. This author admits that the issues identified in this paper have been discussed in the ever-increasing literature on vocational and technical education in Nigeria and elsewhere (Duffy, 1967; Camp, 1983; van Ark, 1992; UNESCO and ILO, 2002; Alwasilah, 2002; Dike, 2006a, 2009a and 2009b, and Maclean and Wilson, 2009). This author is of the view that the analyses in existing studies are more or less superficial and thus limited. The author attempts to conduct an integrated analysis of the issues to change the wrong perceptions the public has about technical and vocational education.

2. Research Questions and Methodology

By reviewing extensively, the existing literature for description and critical analysis of the status, problems and prospects of vocational and technical education in Nigeria, this paper attempts to find answers to the following pertinent questions: Why has Nigeria failed to recognize the importance of technical and vocational education? Can Nigeria transform into an industrialized nation without technological capability? Could India and the "Asian Tigers" have become what they are without proper investment in human development and technical education? (Agrawal *et al.*1995; Mohan, 2003). Based on the studies and the author's critical analysis of the recent development in the sector, this paper brings out that the leaders cannot give vocational and technical education the attention it deserves without changing their 'thinking models' that drive their decisions. Changing their mind-set is a critical first step to restructuring the sector. Without changing their thinking, values and beliefs, it is impossible to change their wrong impression about the field. As it were, poor thinking leads to poor results.

3. Brief History of Vocational and Technical Education

The provision of vocational and technical schools has a long history. As noted earlier, the term has changed over the years because of development in the global economy. Before the Industrial Revolution (between 1750 and 1830) the home and the "apprenticeship system" were the principal sources of vocational education. But societies were later forced by the decline of handwork and specialization of occupational functions to develop institutions of vocational education (Duffy, 1967). Manual training that involves general instruction in the use of hand tools was said to have developed initially in Scandinavia (c.1866). However, vocational education became popular in elementary schools in the United States after 1880 and developed into courses in industrial training, book-keeping, stenography, and allied commercial work in both public and private institutions. As the *Columbia Encyclopedia* (2001) has noted, some of the early private trade schools in the United States include Cooper Union (1859) and Pratt Institute (1887), the Hampton Institute (1868) and Tuskegee Institute (1881). The agricultural high school (1888) of the University of Minnesota was the first regularly established public vocational secondary school that introduced extensive public instruction in agriculture.

The number of public and private vocational schools has greatly increased in the United States since 1900. Vocational education received an impetus during World War II (1939-1945) when the armed services had great need for technicians that the civilian world could not supply. There was a further upsurge on vocational training from the Servicemen's Readjustment Act of 1944 (the G. I. Bill of Rights), which allowed World War II veterans to receive tuition and subsistence assistance during extended vocational training. There was also the Manpower Development Training Act (1962), the Vocational Education Act (1963), and the Vocational Education Amendments (1968) and the Carl D. Perkins Vocational and Applied Technology

Act (1984). These programs help to improve the nation's workforce and ensure that vocational training is available for economically (and physically) challenged youth.

The United States is not the only society that appreciates skills acquired through vocational and technical education. According to van Ark, et al (1992) the Dutch school system pay special attention to high standards in mathematics and the provision of vocational education at ages 14 through 16 for a third of all pupils with widespread vocational education given to students at the age of 16. As the global economy is becoming increasingly driven by information technology, secondary (high) schools in many development-conscious nations are creating vocational and technical centers that offer training in lifelong trade together with general academic studies and vocational and technical schools are shifting their emphasis to training in the computers and information technology.

In Africa, the apprenticeship system was an avenue to acquire vocational skills before the colonial era. And in Nigeria, as in most African societies, the youth were trained in traditional vocations such as pottery, weaving, mat making, wood carving, and traditional medicine, to name but a few, by their parents, family friends, and relatives who were masters of the crafts (Fafunwa, 1974). But everything changed when the colonial masters set up formal vocational schools for those interested in learning particular trades. Early in its history the Royal Society of Arts (RSA) and the City and Guilds of London Institute (CGLI) controlled the craft-level technical education in Nigeria through the conduct of examinations in commercial and technical subjects. But the West African Examinations Council (WAEC) that was created in 1952 took over the conduct of examinations in some technical and commercial subjects from RSA and CGLI, in December 1972. In most cases only the theoretical aspects of the trades were examined by CGLI. Later the federal government introduced the Federal Craft Certificate as practical aspects of the trades examined by CGLI (Oni, 2006; UNEVOC, 1996).

Evidently, the policy makers have not handled Nigeria's quest for development properly as they have failed to give technical education, skills acquisition and general human capital development the attention they deserve. Nigeria's technical institutions' curricula are not structured to meet specific national development needs as the students were given few or no general education to enhance their skills in the preferred trades. Because of the inadequacy of technical education, the graduates were unable to secure admission into formal tertiary institutions. The image of technical education has since suffered as the field is perceived as a program for academically feeble students who are only good at working with tools such as hammer, screwdriver and spanner or wrench. This has apparently contributed to the neglect of technical education in Nigeria. The National Board for Technical Education (NTBE) was established in 1985 for quality assessment and program accreditations at polytechnics, professional institutions, technical colleges and training centers, to set guidelines and standards for admissions to the different programs in each institution. In 1992 however the National Business and Technical Examinations Board (NABTEB) was established and charged with the conduct of technical and business examinations that were conducted by the RSA, CGLI and WAEC (Oni, 2006; UNEVOC, 1996).

Higher levels of technical skills are acquired in tertiary institutions such as the polytechnics, professional institutions (colleges of education), and technical education at the university level. Here most technical education is academic as students are exposed to mostly theory and rudimentary practical application in formal educational settings. The students in two-year programs in polytechnics are awarded a National Diploma (ND) and those in four-year programs are awarded the Higher National Diploma (HND). As noted earlier, because of poor funding the technical institutions lack instructional facilities, workshop and skilled teachers for proper technical education and training. Some of the "half-baked" roadside auto mechanic or other master craftsmen are more skillful in their trade than those who passed through the higher technical institutions. As in other higher institutions the agencies in-charge of technical education in Nigeria have not performed their duties.

The problems facing technical education have a long history. A review of the history of education in Nigeria shows that before independence in 1960, the Ashby Commission was set up to review Nigeria's manpower needs (Ashby, 1960). The Commission reported a shortage of skilled manpower at all levels (lower, middle and high) and observed that "The major defect in Nigerian education is the strong bias toward the traditional literary and academic subjects. This is reflected in a lack of respect on the part of the public for manual labor"-jobs that involve the use of hand tools (Ashby, 1960:18). And the report strongly recommended for Nigeria to introduce technical subjects in secondary education and warned that the nation could "ill afford to ignore" the recommendation. The report also indicated that Nigeria would need to produce hundreds of engineers and thousands of technicians yearly for decades to be able to meet the technical manpower needs of the society.

Although some technical and vocational schools were established following the recommendations of the Ashby Commission, the nation's secondary schools and higher institutions remain oriented toward white collar jobs. This attitude is reflected on the employers' preference for regular university graduates with Bachelors degree (BS/BA) over graduates of Polytechnics with Higher National Diploma (HND) and the pay disparity between the graduates (*This-Day*, 25 February 2008). And the case remains unsettled in spite of government's repeated promises to address the problem. Because the leaders failed to take the Ashby report very seriously, Nigeria is suffering from a shortage of competent lower, middle and high level technical manpower.

4. The Neglect of Technical and Vocational Education

Everyone appears to be wondering why Nigeria has failed to develop as it should have, despite its vast human resources and stupendous oil wealth. Science and technology has been a part of Nigeria's *National Policy on Primary Education* (NPE) since 1981 (Moja, 2000), but like every other policy, its implementation has been poor. While technical and vocational education has continued to thrive in many societies Nigeria has neglected this aspect of education. Every facet of the economy has been affected by the shortage of skilled technicians. The society lacks competent bricklayers, carpenters, painters and auto mechanics, laboratory and pharmacy technicians, electrical/ electronic technicians, and skilled vocational nurses.

And because of the neglect of technical education the hospitals are no longer a place where people go to get their ailments treated, but a place they go and die. Tales abound of how people die out of minor ailments, not to mention major health problems. The refusal of President Umaru Musa Yar'Adua to employ the service of the local hospitals to tackle his chronic health problems gives credence to the general opinion that the local health institutions lack the modern medical infrastructure to handle both minor and major ailments (*BusinessDay,* 27 November 2009). The President has often traveled out to Germany and Saudi Arabia for medical treatment; and other high ranking public officials are also not patronizing local hospitals as well as the educational institutions. The United Nations Children's Fund (UNICEF, 2008) in *The State of the World's Children - 2009* noted Nigeria's poor ranking in maternal, neonatal and infant mortality. According to the report, Nigeria and India account for one third of maternal deaths worldwide. In 2008, Save the Children, a US-based humanitarian organization, claimed that about one million under-5 children die in Nigeria annually. Another study shows that only about 37 per cent of Nigerians have access to good drinking water, and 30 per cent use modern sanitation facilities.

And because of poor training and regulations, the "half-baked" roadside mechanics in the society cause more harm to motor vehicles in servicing them. Although road accident is a global problem, a combination of poor driving training and attitudes toward traffic signs and regulations and poorly maintained vehicles and roads lead to rampant road accidents causing early death of many a people. According to the 2008 report by the *Corps Marshall of the Federal Road Safety Commission* (FRSC) about 4,800 people die in road accidents in Nigeria every year (*Daily Sun*, 20 November 2008). This figure could be conservative, given the sordid state of the roads and menace of armed robbers. The shabby performance of Nigeria's builders (mason/bricklayers) and poor building maintenance often lead to unbridled collapsing of buildings (*ThisDay,* 17 August 2005). For that individuals with important projects today employ competent technicians from neighboring countries (Dike, 2009a). The havoc caused by the poorly trained technicians in the power sector has been well documented as spotty electricity supply is the greatest bottleneck to the growth of the economy and national development (*BusinessDay*, 10 July 2007).

The agricultural sector too is a victim of the shortage of technical manpower as toiling all day in the field with knives, hoes and shovels would not feed the nation's 140 million people. Because of the shortage of mechanized farming, which requires technical skills, Nigeria imports billions of naira worth of food annually to feed its population (*BusinessDay,* 19 November 2007). Even the financial sector is not spared as it lacks competent technicians to establish databases and credit bureaus to ascertain their customer's credit worthiness and true identity. Because of this the banks often give out huge loans to individuals who otherwise would not qualify. The sector also lacks technical expertise to regulate and supervise the banks and to develop financial software to properly tackle the rising fraudulent activities in the banking sector. Thus because of the sector's systemic weakness some bank managers and other top officials connive with crooks (politicians and business executives) to defraud their banks. All these contributed to the huge non-performing loan crises that rocked some banks in recent times (*Daily Independent*, 17 November 2009).

Because of poor education and training and motivation the police extort money as low as N20 from the public and violate the human and civil rights of the citizens who question their authority. The police also lack skilled forensic laboratory and fingerprint technicians to conduct criminal investigations. There was a shameful episode recently in the society where the police paraded a goat as a thief; the police noted that the criminal they were chasing transformed into an animal (*BBC News*, 23 January 2009). There is a widespread public concern about the danger posed by environmental pollution and fake drugs in the society. The *National Agency for Foods, Drugs Administration and Control* (NAFDAC), the agency responsible for controlling the authenticity of foods and drugs lacks competent technicians to properly monitor the manufacturing, import and distribution of fake/expired drugs in the society (*Leadership*, 27 September 2009). The poor and less educated individuals lack the skills and knowledge to manage their health problems such as AIDS, cancer, diabetes and high blood pressure.

The neglect of technical education is socially and economically injurious because it is robbing the nation of the contributions the graduates would make on national development. As this author has noted elsewhere (Dike, 2006a) the under-development status of Nigeria could be linked to the odious neglect of its educational institutions (particularly technical education) and investment in other critical areas of human capital development such as health care and provision of social amenities. For Streeten (1984) development of human capital would help any nation achieve to some extent 'self-sufficiency in food production, capital, and goods and services' and 'the understanding of the nature of the environment, the preservation of it and eventually will eradicate environmental degradation, desertification, deforestation and soil erosion.'

Most of the so-called "expatriate engineers" who are being paid millions of dollars to build Nigeria's roads and bridges are graduates of technical and vocational colleges. Yet Nigeria does not take its vocational and technical institutions seriously. Although technical and vocational education seem deficient in 'citizenship or leadership training' (Friedman, 1982), it provides, however, the youth with "life skills" (Alwasilah, 2002) to enable them become productive entrepreneurs, engenders creative and innovative minds, and enlarge the economic pie, and thus, increase personal freedom. As Sen (1999:36-37) has aptly noted, "Freedoms are not only the primary ends of development, they are also among its primary means."

Poor leadership and meager spending on education are among the problems facing Nigeria's educational institutions (*Daily Sun*, 27 October 2009). Thus 'Leadership without a moral purpose' (Dike, 2009c) has subjected the citizens to immeasurable economic hardship and misery as the political leaders are not doing enough to improve the people's living conditions and affected teachers' morale and productivity. A review of available data on African nations spending (per cent of GDP) on education shows Nigeria's unimpressive position.

It has vastly been documented that public disdain for technical education is the major reason for the odious neglect and low enrolment in technical colleges in relation to secondary school enrolment (see below). Because of paucity of data current enrolment figures in technical colleges are not available. Nevertheless, if the 2001 figure is used as sample the number of enrolment in technical colleges from 2002 to 2009 would not be impressive.

Notwithstanding the need for more science and technology graduates or the need for vocational and technical education, if you will, the government has continued to award licenses to organizations and individuals to establish regular private universities, which are poorly equipped. Consequently, many of the graduates lack employability skills that could be acquired from a well-equipped two-year technical college. Because the local higher institutions are unable to meet the demand for advanced technical manpower, employers in the public and private sectors are forced to hire technicians from technical and vocational institutions abroad to the chagrin of the public (Guardian, 17 August 2007). Experience shows that the massive layoffs resulting from the on-going global economic crisis have affected the youth and others without good education, job training and technical skills disproportionately.

Table 1: Selected African Nations: Education Spending (Per cent of GDP)

Country	% GDP	Country	% GDP
Angola	2.7	Mozambique	2.4
Algeria	4.3	Namibia	8.6
Cote d'lvoire	4.6	Nigeria	1.3
Cameroon	3.2	Rwanda	2.8
Ethiopia	4.8	Senegal	3.2
Eritrea	4.8	Sierra Leone	2.7
Egypt	4.5	South Africa	5.7
Kenya	6.4	Sudan	1.4
Liberia	4.5	Tunisia	6.8
Libya	3.8	Uganda	2.3
Madagascar	3.2	Zambia	1.9
Mali	2.8	Zimbabwe	4.7
Morocco	5.5		

Sources: Compiled by the author: Data are extracted from High School Social Studies/Geography manual used by the Sacramento Unified School District, Sacramento, California (2009-2010); and Dike, V.E. (2009c), *Leadership without a Moral Purpose: a Critical Analysis of Nigerian Politics and Administration* (with emphasis on the Obasanjo Administration, 2003-2007), North Charleston, South Carolina: Book Surge Publication.

Table 2: Enrolment in Secondary and Vocational &Technical Institutions, 1991-2001

Year	Number of Technical Colleges	Student Enrollment	Number of Secondary Schools	Student Enrolment
1991	208	46,083	3,854	1,653,891
1992	202	40,878	5,840	1,814,000
1993	190	72,136	5,948	1,865,189
1994	300	72,136	6,092	2,794,498
1995	240	76,434	5,991	2,934,349
1996	252	89,536	5,859	2,941,781
1997	261	1,179	6,001	2,923,791
1998	261	1,426	5,860	2,901,993
1999	261	1,425	6,008	3,123,277
2000	261	1,835	6,009	3,600,204
2001	261	1,835	5,959	4,032,083

Sources: Compiled by the author from Social Statistics in Nigeria 1995; Annual Abstracts of Statistics 2001 (Federal Office of Statistics 2001), Nigeria; Oni C. S. (2006), 'Vocationalism in Nigerian Education,' in Journal of Social Science, 12(2): 147-150.

5. The Problem of Youth Unemployment

As noted earlier, Nigeria's current preoccupation with university education reduces economic opportunities for those who are more oriented toward work than academe, thus leading to rising youth unemployment. More often than not the public perceives students who take to vocational and technical education as those who lack the ability to continue with formal academic studies in higher institutions. But not everyone needs a university education. Sadly, in

Nigeria social recognition and advancement on workplace depend to a large extent on the stack of academic degrees one has acquired. Because of poor public perceptions about blue-collar jobs the youth are not enthusiastic about technical and vocational education. The few technical and vocational schools in the society also lack qualified teachers and functional workshops and equipment for hands-on application.

Youth unemployment has been skyrocketing in Nigeria because they lack the skills employers need as well as for self employment. The federal government recently acknowledged that about 80 per cent of Nigeria's youth is unemployed and 10 per cent underemployed (*Daily Trust*, 26 November 2008). Others have urged the youth to become "entrepreneurs" and reject "social vices" (*This-Day*, 17 March 2009). At some point the Minister of Education, Sam Egwu, has expressed concern about the poor quality of graduates from the nation's educational institutions. The political leaders are good at cataloging the reasons for Nigeria's problems and predicting the future without implementing reasoned policies to create a better future. But as Peter Drucker has observed "If you want to predict the future, create it" (as cited in Wilson and Blumenthal, 2008:1) It is not good enough to implore the youth to become productive citizens and "entrepreneurs" or to reject "social vices" without providing them with skills and resources.

6. Recent Policy Development

Recent major policy development of the Yar'Ardua administration revolves around his unrealistic aspiration of transforming Nigeria into one of the first 20 largest global economies by the year 2020 and the Seven-Point Agenda. And the drumbeat has been rising to a deafening height as he recently assembled a 405-member panel to realize the Vision 2020 project (*Daily Trust*, 16 February 2009). The political leaders can make all the noise they want about the impracticable policies of transforming Nigeria into an industrialized nation, but the fact remains that rhetoric alone cannot make the nation an economic super-power. But the Vision 2020 program and the Seven-point Agenda will remain a paper tiger without effective institutions and infrastructure and technological capability. History shows that no society has become an industrialized nation without technological capability (Mohan, 2003). This means that for Nigeria to become an industrialized nation copious investment in human capital development (general education and science and technology, i.e. technical and vocational education in particular) must be a major part of any development strategy. Any person who thinks that a country that lacks the skilled technical manpower and infrastructure to generate and distribute electricity for more than three hours in a day and to maintain its roads and bridges could be transformed into an industrialized nation in less than eleven years from today must be living in a different planet.

Another bogus policy is the re-branding campaign being trumpeted by the Information and Communications minister, Dora Akunyili and the buy 'made-in-Nigeria goods and services' crusade of Achike Udenwa, Minister of Commerce and Industry (*Daily Independent*, 23 October 2009). Meanwhile, Dora Akunyili tried to rationalize the N150 million earmarked for the project in the 2009 budget by arguing before the joint National Assembly Committee on Information that N1.49 billion was spent on a similar re-branding project -the 'Heart of Africa' - during the Obasanjo presidency (1999-2007).

The funds being wasted on all the impracticable projects could have been wisely invested on public infrastructure and education to grow the economy, create jobs and stabilize the polity. The re-branding campaigns cannot eradicate corruption and fix the erratic power supply and the pot-holed roads. Neither will it clean up the heap of trash littering the streets or change the mentality of the leaders. The campaigns will not persuade the world to recognize Nigeria as a "Great Nation" or to buy the poor quality made-in-Nigeria goods and services. As George Washington Carver says, "There is no shortcut to achievement. Life requires thorough preparation."[1] Nigeria can command the attention of the international community only with good governance and high quality goods and services. This means that Nigeria needs a new paradigm for effective transformation of its economy and educational institutions--particularly vocational and technical education.

7. Vocational and Technical Education in Nigeria's Development Strategy: A New Paradigm for Effective Transformation

Having traced the trends and state of vocational and technical education this section will consider some of the practical ways to integrate this sector in Nigeria's development strategy. As mentioned earlier, it has been well documented that vocational and technical education has a low public status (Moja, 2000). But despite the demand for advanced technical manpower the policymakers have failed to adopt appropriate policy to reverse this trend. It has been noted that a nation's economic prosperity, political influence and power in the global arena depend precariously on its capability to employ science and technology for national development. Any nation that wishes to remain politically relevant must strive to raise the standard of living of the people through the employment of advanced scientific and technological knowledge (Okoye, 1999). And that means investing in human capital development.

Advanced technical skills and knowledge would enable Nigeria to maximize the exploitation and utilization of all available natural resources, mechanization of the agricultural sector and effectively and efficiently transform raw materials into goods and services, create jobs and bring about the availability of good water supply for domestic and industrial use. And that means good quality of life for the people. Because of the importance of vocational and technical education the government should adopt new and better policies to address the deficiencies in the sector and properly integrate it in national development strategies.

Soaring youth unemployment leads to poverty and social crisis; and like unemployment, poverty is ravaging the society (Dike, 2003a). The *World Development Indicators* published by the World Bank in 2007 shows that more than 70 per cent of Nigerians are living on less than $1 per day. (Reportedly, the World Bank changed the poverty line to living below $1.25 a day in 2008.) Skills acquisition is one of the tried-and-true ways to reduce the nation's high rate of joblessness and poverty. However, Nigeria's poverty alleviation programs have been ineffective because of poor education, lack of skills training, lack of public assistance to the needy, and corruption. Throwing some money to the poor and illiterate citizens who cannot manage their own lives to set up small business is like pouring water in a bucket with holes.

The design of Nigeria's educational curriculum is flawed given the neglect of technical and vocational education. Consequently, the nation is today lagging behind in preparing its labor force for the 21st century economy. As a policy, the high (secondary) schools should be required to establish technical and vocational departments (or centers) where the youth could learn life-long trade and to solve authentic technical problems with mechanical skills. In addition, there should be some form of school-work-based learning incorporated in schools as integral part of national development strategy (Dike, 2006a). Specifically, the schools should provide real-world examples on how the students would apply what they learned and in-class instruction should be relevant to the student's needs (Camp, 1982). Toward these teachers must learn to blend theory with practice because theories alone cannot serve any useful purpose.

The nation's industrial productivity has been declining in recent years (*Daily Independent*, 06 November 2007) because of a combination of bad business environment, lack of skills, poor reward systems, and low workers' morale (Dike, 2003b and 2003c). Empowering the people with relevant job skills would enhance their productivity and national development because the progress of any society lies on the quality of its education, institutions and infrastructure, and workers' productivity (Lewis, 2004). Higher productivity gives a nation the advantage of economies of scale, lowers the costs of production and prices of goods and services, and improves its global competitiveness (Dike, 2006b and 2006c).

One of the ways to motivate workers and improve their productivity is to give them the skills and knowledge they need to effectively perform their duties (Argyris, 1994; Pintrich and Schunk, 1996). Granted that Nigeria's workers are subjected to appalling working conditions, but constantly calling them out for industrial actions is not the only way to fight for their welfare (*Nigerian Tribune*, 26 October 2009). The Nigeria Labor Congress (NLC) and the affiliated unions should establish technical and vocational training centers in the local government areas to impart the workers' employability skills (Dike, 2003b and 2003b). In today's crisis-ridden economy one of the ways to spur the weak economy is to empower the people with relevant skills to tackle their personal problems and the challenges facing the nation. Thus the Academic Staff Union of Polytechnics (ASUP), the Academic Staff Union of Nigerian Universities (ASUU), the National Board for Technical Education (NBTE) and teachers in this area should seriously campaign for increased funding for vocational and technical education and launder its image as part of the current agitation to improve the state of education.

To move forward the society should discard its wrong perception on technical and vocational education. Adult education is another aspect of non-formal education that has been neglected. As in technical and vocational education, the consequences of the neglect of adult education are glaring (Nnazor, 2005) because no nation would make any meaningful socioeconomic stride without the citizens being literate to a certain degree. In advanced nations individuals with technical skills and field experience in relevant fields are highly respected and thus work in tandem with those with formal education. The worth of every worker should depend on the person's skills and knowledge and not the stack of academic degrees one has. Secondary and tertiary education should encompass economic, educational and social objectives because workers need more than academic knowledge to function well on the job. For any worker to be productive and compete (Hamel and Prahalad, 1994) effectively in the rapidly evolving knowledge-driven global economy the worker must possess relevant job competencies, including technical, business, cultural, interpersonal and intellectual competencies. All these could be obtained in well-equipped technical and vocational colleges.

The nation's technical colleges and schools should be properly equipped and teachers with skills and knowledge in the subjects should be employed and motivated to properly educate their students; and professional technical educators (and not politicians) should be hired to manage the technical institutions. To determine their competencies and to ensure standards, the graduates of technical institutions should be thoroughly tested, certified and registered before permitting them to work as technicians in their chosen fields. And the agencies in-charge should revoke the certificate of any technician whose work fails to meet the stipulated standards. Technical and vocational education holds the key to Nigeria's development (Dike, 2009a). No nation can fight a war without an army. In the same token Nigeria cannot become an industrialized nation without technological capability.

8. Concluding Remarks

It is the responsibility of every government to provide the citizens with the critical skills and resources to realize their full potential. To move forward Nigeria must take investment in human capacity development (education and skill training) very seriously as no nation can compete effectively in the emerging computerized global market place with poorly educated and unskilled workers. Nigeria needs to adopt a skill-oriented educational principle because how quickly a nation recovers from the present global economic predicament depends on the strength of its institutions and infrastructure, ingenuity of the leaders, and productivity of workers (Dike, 2009d). Any administration that can give vocational and technical education the attention it deserves, stimulate the weak economy, and create jobs for the army of unemployed youth, will occupy an enviable spot in the nation's political and economic history. Scarcity of jobs combined with lack of relevant job skills and knowledge causes higher degree of youth unemployment.

The government should not be expected to tackle the problems facing technical education alone. Although UNESCO has assisted Nigeria over the years in tackling the problems facing vocational and technical education, more assistance is needed from international partners, the private sector and NGOs and other stakeholders. The policy makers should implement reasoned policies to deal with the debilitating deficiencies in the education sector instead of wasting time and resource on the false re-branding campaigns. 'Visions' and 'Re-brandings' will not solve Nigeria's socio-political and economic problems. The leaders could salvage Nigeria's image by re-branding their outlook to do the right thing. The choices we make have a great impact of the quality of our lives. Without a fundamental shift in values, beliefs and thinking, and without appropriate investment in human capital development and without technological capability, Nigeria will continue to only dream of becoming an industrialized nation.

Notes

1 George Washington Carver (Quote): as cited in the school poster (no date) on the wall in the Main Office at George Washington Carver School of Arts and Science (Charter School), Sacramento City Unified School District; 10101 Systems Parkway, Sacramento, CA 95827.

References

Agrawal P., Gokarn S. V., Mishra V., Parikh K. S., and Sen K. (1995), *Economic Restructuring in East Asia and India, Perspectives on Policy Reforms*, New Delhi, India: Macmillan Ltd.

Alwasilah C.A. (2002), 'Vocational education must provide students with life skills,' in *The Jakarta Post*, February 11.

Argyris C. (1994), 'The future of workplace learning and performance,' in *Training and Development*, May, 48(5), pp. 36-47.

Asche M. (1991), 'Educational Reform and Vocational Education: Review with Implications for Research and Development,' in *The Journal of Vocational Education Research* 16 (3):1 – 34.

Ashby E. (1960), *Investment in Education: The Report of the Commission on Post School Certificate and Higher Education in Nigeria.* Lagos: Federal Ministry of Education, p. 18.

BBC News (Africa) (2009), 'Nigeria police hold 'robber' goat,' January 23.

BusinessDay (2007), 'Why Nigeria imports N381bn food annually,' November 19.

BusinessDay (2009), 'Only 15 surgeons in Nigeria can treat Yar'Adua's ailment,' November 27.

Camp W. G. (1983), 'Social efficiency and vocational education: An examination of our changing philosophies,' in *Journal of Vocational Education Research*, 8(3), pp. 10-19.

Camp W. G. (1982), 'Social efficiency revisited: A cornerstone of our foundation,' in *The Journal of the American Association of Teacher Educators in Agriculture*, 20(3), pp.11-18.

Columbia Encyclopedia (2001) (6th edition), New York: Columbia University Press.

Daily Independent (2009), 'CBN Bank-debtor List Renews Urgency for Credit Bureaus in Nigeria,' November 17.

Daily Independent (2007), 'Economy operates at 25% capacity-Soludo,' November 6.

Daily Independent (2009), 'Udenwa Urges Nigerians to buy Made-in-Nigeria Goods', October 23.

Daily Sun (2008),'Nigeria's grim road accident statistics,' November 20.

Daily Sun (2009), 'Poor Leadership, bane of tertiary education-Pro-Chancellor,' October 27.

Daily Trust (2008), '80% of Nigerian Youths Unemployed-FG,' November 26.

Daily Trust (2009), 'Yar'Adua inaugurates Vision 2020 business support group today,' February 16.

Dike V.E. (2003a), 'The Global Economy and Poverty in Nigeria,' in *NESG-Economic Indicators*, 9(1):7-16.

Dike V.E. (2003b), 'Reward system and labour productivity,' *Daily Independent,* May 15.

Dike V.E. (2003c), 'Toward labour productivity,' *Daily Independent,* May 22.

Dike V.E. (2006a), 'Youth Unemployment in Nigeria: The Relevance of Vocational and Technical Education,' in *NESG -Economic Indicators*, 12(3): 25-29.

Dike V.E. (2006b), 'Education and Global Competitiveness (1),'*Daily Independent*, November 17.

Dike V.E. (2006c), 'Education and Global Competitiveness (11),' *Daily Independent*, November 18.

Dike V. E. (2009a), 'Technical and Vocational Education: Key to Nigeria's Development (1),' *Daily Triumph*, March 27.

Dike V. E. (2009b), 'Technical and vocational education: Key to Nigeria's development (II),' *Daily Triumph*, March 30.

Dike V. E. (2009c), *Leadership without a Moral Purpose: a Critical Analysis of Nigerian Politics and Administration* (with emphasis on the Obasanjo Administration, 2003-2007), North Charleston, SC: BookSurge Publication.

Dike V.E. (2009d), 'Global Economic Crisis and Power of Productivity,' in *NESG -Economic & Policy Review,* 15(2): 27-32.

Doll R. and Peto J. (1985), *Asbestos: Effects on Health of Exposure to Asbestos*. London, HMSO, 1985.

Drucker P. (Quote) as cited in Wilson J.S. and Blumenthal I. (2008), 'Managing Brand You: 7 Steps to Creating Your Most Successful Self,' American Management Association (see Introduction: What Branding Can Do for You), p.1.

Duffy N. F. (ed.) (1967), *Essays on Apprenticeship*, Wisconsin: Center for Studies in Vocational and Technical Education.

Fafunwa A.B. (1974), *A History of Nigerian Higher Education,* Lagos, Nigeria: Macmillan Publisher.

Friedman M. (1982), *Capitalism and Freedom,* 2nd edition, Chicago: University of Chicago Press.

Gleeson, D. and Mardle, G. (1980), *Further Education or Training? A Case Study in the Theory and Practice of Day Release Education*, London: Routledge & Kegan Paul.

Guardian (2007), 'Don't blame employers of labour,' August 17.

Hamel G. and Prahalad C.K. (1994), *Competing for the Future*, Boston: MA: Harvard Business Press.

Leadership (2009),'NAFDAC Losing War against Fake Drugs-Pharmacists,' September 27.

Lewis W.W. (2004), *The Power of Productivity: Wealth, Poverty, and the Threat to Global Stability*; Chicago: University of Chicago Press.

McNabb J. G. (1997), 'Key affective behaviors of students as identified by a select group of secondary school teachers using the SCANS categories,' in *Journal of Industrial Teacher Education*, [On-line edition] 34(4). See: http://scholar.lib.vt.edu/ejournals/JITE/v34n4/html/mcnabb.html (accessed December 2009).

Mohan R. (2003), *Facets of the Indian Economy: The NCAER Golden Jubilee Lectures*, New Delhi, India: Oxford University Press.

Maclean R. and Wilson D.N. (eds.) (2009), *International Handbook of Education for the Changing World of Work: Bridging Academic and Vocational Learning*, New York, NY & Bonn, Germany: Springer.

Moja T. (2000), 'Nigeria Education Sector Analysis: an Analytical Synthesis of Performance and Main Issues,' *World Bank Report 2000*, Washington D.C.: World Bank.

Nigerian Tribune (2009), 'ASUU strike: The costs, pains, gains,' October 26.

Nnazor R. (2005), 'Adult education in Nigeria: The consequence of neglect and agenda for action,' in *International Education Journal*, 6(4): 530-536.

Okoye C.M. (1999), 'The Impact of Science and Technology on Nigerian Development,' in *Journal of Science Education,* 6 (1): 20-29.

Oni C.S. (2006), 'Vocationalism in Nigerian Education,' in *Journal of Social Science*, 12(2): 147-150.

Pintrich P. R. and Schunk D. H. (1996), *Motivation in Education: Theory, research, and applications*. Englewood Cliffs, NJ: Prentice Hall.

Sen A. (1999), *Development as Freedom,* New York: Anchor Books (see Chapters 1-3), pp.36-37.

Streeten P. (1984), 'Human Development: Means and Ends,' *The American Economic Review: Papers and Proceedings of the Hundred and Sixth Annual meeting of the American Economic Association*, January 3-5, Boston, M.A..

ThisDay (2005), 'When Buildings Collapse,' August 17.

ThisDay (2009), 'Shun Social Vices, Youths Told,' March 17.

ThisDay (2008),'FG to Address disparity between Degree, HND Holders,' February 25.

UNESCO and ILO (2002), *Technical and Vocational Education and Training for the Twenty-first Century- UNESCO & ILO Recommendations.* Available at: http://unesdoc.unesco.org/images/0012/001260/126050e.pdf (accessed December 2009); Paris, France: UNESCO; Geneva, Switzerland: ILO.

UNEVOC (1996), 'Technical and Vocational Education in Nigeria,' in *The Development of Technical and Vocational Education in Africa: Case Studies from Selected Countries,* Chapter 7. p. 410.

UNICEF (United Nations Children's Fund) (2008), The *State of the World's Children 2009: Maternal and Newborn Health* (Series - State of the World's Children), New York: UNICEF.

Van Ark B., Mason G., Prais S.J. (1992), 'Vocational education and productivity in the Netherlands and Britain,' in *National Institute Economic Review*, 140(1): 45-63.

Warnat W.I. (1991), 'Preparing a Worldclass Workforce.' in *Vocational Education Journal* 66 (5):23- 25.

Washington-Office of Superintendent of Public Instruction (no date), 'Career and Technical Education,' Available at: http://www.k12.wa.us/CareerTechEd/ (accessed November 2009).

Wirth A.G. (1972), *Education in the Technological Society: The vocational-liberal studies controversy in the early twentieth century*, Scranton, PA: Intext Educational Publishers.

CHAPTER 3

Reaching Out:

Efforts to Build Sustainable African Research and Innovation Capacity

Joseph M. Kizza, Mammo Muchie & Timothy Waema

1. Introduction

Research and innovation are critical for building productive power and the creation of wealth. It is even more critical in the African environment where currently the culture of both research and innovation is minimal. Currently there are still many African countries that cannot sustain at least one good research university. Having at least one institution that carries out both basic and applied research in relevant areas of development would create a badly needed engine for developing the knowledge base essential for that country's economic growth and social development. So far this is not the case. As Damtew Teferra,[1] Founding Director, International Network for Higher Education in Africa (INHEA), observes, the levels and quality of research, analysis, and communication about higher education in Africa still remain woefully underdeveloped. There is still lack of basic accessible data, in depth analysis, and a central hub for information on the sector. There is still a high magnitude of the knowledge gap in this area. In fact, there are many still arguing against "flagship" universities in favor of only expanding "community college" like institutions in Africa. At a time when knowledge has become a global currency for social and economic progress and universities globally are striving to raise their competitiveness, such recommendations are both curious and down right destructive, according to Teferra:[2] Research and innovation capacity based on strong research university is core component of any country's economic, political and social development systems. Ways must be found to create the needed research and innovative capacities starting with the setting up of African based research institutions including universities and academies. The doubts about the need to set up research universities in Africa need to end and instead ways of making them effective pillars for building Africa's productive and innovative capabilities must be promoted with an acute awareness of much squandered time by debating whether universities are necessary instead of building them as part of the core producer of Africa's innovation systems.

The Context for Action

African universities are still emerging from the difficulties they faced from the 1970s through the 1980s to end of the 1990s. The American Chronicle of Higher Education recently stated that, "African universities' capacity to educate new PhD holders is eroding, raising deep concerns about the continent's ability to produce new generations of academics and educators".[3] Between 2000 through 2004, Africa produced only 1.8 % of the world's publications. In comparison India produced 2.4% and Latin America 3.5% of the world's research.[4] Moreover it has been suggested that much of the African research is concentrated in only two African countries, i.e. South Africa and Egypt which between them account for just over 50% of the continent's publications. The top 8 countries in Africa produce above 80% of the continent's research.[5] Africa's inventive profile also shows that the continent needs to catch up; for it produces less than one thousand of the world's inventions, and where 88% of the continent's inventive activity is located in South Africa.[6]

Invariably when it comes to knowledge, research and higher education in Africa, the picture that emerges is unflattering regarding research and training pertaining to all levels from the local, regional to the continent; the reasons for this are many including:

(a) Weak investment from national governments to promote continental knowledge infrastructure

(b) Weak mobility of knowledge, research, researchers, curriculum and innovation horizontally

(c) Brain drain to the Northern hemisphere. Brain migration is a real problem given the lack of brain circulation within the continent.[7]

(d) Migration of good quality staff and researchers to the private sector also drains human capital away from education and research.

(e) Obstacles and lack of incentives to stimulate inter-African knowledge communication

(f) Poor or no representation in national development plans and bilateral donor support.

(g) Continuing 'competition' with universal primary education for popularity and extremely limited resources.

(h) Low reproduction rates. Increasing age profile of top researches with retirement of senior and experienced staff combined with decreasing candidates for replacement.[8]

(i) National and regional needs for graduates far outweigh current higher education institutional capacity.

(j) Low interest and too few graduates in science and technology.

(k) Poor levels of quality research and publications.

(l) Poor communications and transport infrastructure.

(m) Cross-border regulatory obstacles to regional collaboration.

(n) High dependence on international donors for the foreseeable future.[9]

(o) On the positive side there is now a clear recognition and a desire to promote research and innovation, higher education and knowledge across Africa. However, the reality on the ground for African universities is still difficult with a lack of concentration of talent, resources and acute governance crises. In addition, as late as last November reports suggest that African universities still face a looming shortage of PhDs.[10]

The importance of building research training in Africa

Research is not a luxury,[11] but is rather a productive power, pivotal to the development of national systems of innovation and overall economic development[12]. This is clearly recognised by the African Union and NEPAD which have recommended that 1% of GDP by each African state should be spent on R&D. The Ministerial Forum on Health in its meeting in November 2008 formally recommended 2% of GDP on health systems research as part of national research systems development in relation with other systems. Countries like Botswana plan to upgrade their R&D for the next five years.

South Africa has the strongest research base on the continent. The South African Department of Science and Technology has designed a ten-year innovation plan (2008-2018) where the national leadership has understood the importance of research and knowledge. Since 1994 the country has managed to double the science budget. The South African government plans to create 210 research chairs by 2010 and 500 by 2018. It has plans to create 600 PhDs, with 300 in science, technology and engineering by 2018. It hopes to increase its global share of publications output by 1.5 % from its 0.5 % in 2006. Patent applications are expected to grow from 418 in 2004 to 2,100 by 2018. South Africa's own patent office expects to grow from 4,721 in 2002 to 24,000 in 2018. Currently, South Africa has five research universities recognised by international rating and 8 Nobel laureates.[13]

2. African Research and Innovation Capacity Building Initiatives

The challenge is not only to find ways to link these South African research initiatives with other African countries, but also to come up with new initatives to rejuvenate research and innovation broadly across Africa. In this pursuit several initiatives need to be explored. Among these are the involvement of a large African diaspora, creation of local private-public networks to drive African research and innovation. Examples of such networks include the Triple Helix Association and NESGLOBAL[14] to stimulate knowledge remittance back to Ethiopia.

In the rest of this paper, we shall discuss some of the initiatives the authors have initiated or are directly involved in to boost capacity in research an innovation.

Log In Africa (2005 - 2008)

LOG-IN Africa) was an emergent Pan-African network of researchers and research institutions from eight countries – Egypt, South Africa, Morocco, Ethiopia, Kenya, Uganda, Mauritius and Mozambique. Funded by International Development Research Council (IDRC), the network was intended to assess the then prevailing state and outcomes of e-local governance initiatives in Africa, and in particular how ICTs ware being used to realise good local governance at four levels:

- the internal organizational processes of local governments;
- the provision of information and service delivery;
- the promotion of the principles of good governance; and
- public participation and consultation.

With Prof. Tim Waema as the Network Coordinator and Prof. Joseph Kizza and Prof. Mammo Muchie as scientific advisors to the researchers, the network started in 2005 and ended in 2008 with high achievements.

Achievements of Log-In Africa Network

One of the achievements of this study was a roadmap to guide the implementation of e-governance projects in local governments (Waema, 2009). The roadmap was informed by a framework derived from strategic management literature with four key questions that a local government must answer. The first question was: Where are we? This question sought to assess the context of e-Governance both at the national and local government levels. The second question was: Where do we want to go? With this question, a local government must be crystal clear of its vision for e-Local governance, or the desired end-state in the longer term. Obviously this vision must be aligned to the general vision for the local government. In the roadmap developed, we recommended such a vision as:

> "e-Government implementation resulting in good local governance outcomes and contributing to local economic and social development".

The next question was HOW do we get there? With this question, the local authority was seeking to develop a set of appropriate strategies of how to address the key issues from the assessment of e-Governance context (question 1) and to realize the vision (question 2). In the roadmap developed, strategies were developed to address for the following thematic areas (Waema, 2009a, 2009b):

(i) Alignment with governance policy and e-Governance strategy

(ii) Creation of political will

(iii) Communication

(iv) Development of ICT infrastructure

(v) Development of technical, managerial and user ICT human resource capacities

(vi) ICT funding

(vii) Affordability of e-Local government services

(viii) e-Local governance business processes

(ix) e-Local governance project management

The question of how to get there also concerned itself with the methodologies and tools that are useful in the implementation of e-Local governance projects. The roadmap focused on the use of business process mapping and e-Local governance applications design methodologies and tools. This was largely because the research established that e-Local governance applications were implemented without documenting the business processes and re-engineering them for increased efficiency. In this question, the roadmap also focused on the use of a proven methodology for e-Local governance implementation processes. This methodology was meant to guide the local government in the roll out of e-governance projects in order to increase the success of implementation.

The Log-In African Research Network came to an end in September 2008 and achieved the following:

- Established an e-Local governance research network (www.loginafrica.net).
- Successfully completed eight projects on e-Local governance.
- developed an e-Governance assessment framework, which has been published to shape research in this area.
- Synthesised all the findings into an e-local governance roadmap. This is being used to inform African governments in e-governance policy formulation, enabling e-governance policy re-orientation and providing best-practice guidelines in e-local governance project implementation.
- Developed Indicators for e-governance outcomes which can be used to plan, monitor and evaluate e-governance projects. They are also useful in establishing the link between ICT and governance.
- Published over 30 papers in refereed journals and conferences/ workshop proceedings.
- publishing a book on the research findings. This book is a valuable repository of nine rich cases of e-Local governance in Africa.

- Organised a successful international conference on e-local governance. The proceedings of this conference were published by Editions Marocaine et Internationale (EMI), Tangier, Morocco.
- Developed and tested e-government business process mapping methodology and prototype software by the Egyptian team, the citizen registration system by the Moroccan team and the prototype system for automation of life event services by the Ethiopian team.
- Increased and reinforced institutional capacity in ICTs for local governance in both central and local governments in a number of African countries. In addition, the LOG-IN Africa network increased research capacity on e-Governance in participating research institutions.
- Raised awareness of the role of ICTs in local governance was significantly raised among policy makers and practitioners.

ICT for Development (ICT4D)

This initiative funded by IDRC and was led by Prof. Timothy Waema, with Prof. Mammo Muchie and Joseph Kizza acting as mentors and resource personnel in the training of young African researchers.

Objectives

The aim of the initiative was to fund and enable Masters and PhD students in universities in Eastern and Southern Africa countries to carry out research on different aspects of ICTs and development. It was hoped that this would help the regions in:

- Building institutional research capacity on ICTs and development in the various universities in the region and this would lead to the creation of knowledge on the relationships between ICTs and development and the associated impacts.
- Building human capacity on ICTs and development in the region.
- Increasing intra-Africa partnerships and collaborations. The University of Nairobi, like many other African country univer-

sities, has limited relationships with other African universities in comparison to the high number of linkages with European and North American universities.

As part of this initiative, the authors were involved in running a six-day workshop attended by over 20 young researchers from several African universities with the following objectives:

- To help students to identify the research problems and clearly define them in their project proposals.
- To prepare and equip grantees for the entire research process, from data collection to its analysis, both quantitative and qualitative.
- To endow grantees with the knowledge to write proposals and basic skills for grant management.
- To instruct the students in the writing and publishing of scientific papers.
- To train students in the basic skills of making oral presentations in conferences and workshops.
- To offer a forum for networking with fellow researchers from all over Africa.
- To provide for elaborate evaluation of the grantees' projects at their various stages.

An African Globelics Academy or Doctoral School

To meet the challenge of reversing the decline of research and a research culture in Africa experienced over a generation, the authors have embarked on starting an African Globelics Academy for Research, Innovation and Capability building (AGARIC). The AGARIC is intended to be a creative doctoral training and quality enhancing mechanism to create the best quality emergent researcher in Africa. It will involve two areas of research co-joined, namely Science, Technology and Innovation for Development (STI4D) and Information Communication Technology for Development (ICT4D), and two types of academic mentors will share the platform for two weeks to undertake the most intensive PhD training in Africa. The main objective is to foster, sustain and embed a research and innovative culture where those in engineering network naturally from every part of Africa with those coming from social and human science. Staff and PhD candidates will be drawn from both engineering, computer science and social sciences. Staff and candidates will also be included from the rest of the world to support and strengthen the research culture and the African research agenda.

The key reasons for organizing AGARIC can be summarized as:

- The need in Africa to develop and organize a strong research community andcreate an African research culture
- The importance of linking the relatively strong research in South Africa with the rest of Africa
- The significance of drawing on the best from the rest of the world to support the creation of a vital African research and innovation culture
- The need to break the silos in research and innovation by creatively designing and sustaining implementation directed African networks that can draw researcher interactions spanning from human and social sciences to the applied and basic engineering and physical sciences.
- The importance of flexibility in designing and training doctoral academies especially in Africa to make sure what is critical is delivery and results and not building bureaucratic authority that impede implementation and also to support universities and not create independent research NGOs.
- The importance of building AGARIC for all the sciences where at any time the training should involve at least one technical subject area, and another social area. In addition, socio-technical areas that combine both technical and social (e.g.: systems of innovation, development economics of innovation) with ICTs (e.g., from the LOG-INAfrica research network)
- The Camel model where mobility is encouraged across fields, universities, academies and regions will be stressed.
- The creation of an Africa Research community by breaking the Luso-phone, Francophone, Anglophone, Levantine phone tradi-

tions to create ease of mobility and interaction will be acknowledged

- The need to strengthen the Africa research area to encourage and attract the best from global trends of knowledge, research and scientific developments will be stressed.
-

Objectives

The long term objectives of the AGARIC project may be summed up as follows:

- Establish an annual training programme of PhD researchers working on various aspects of science, technology and innovation for development (STI4D) and Information, Communication Technologies for Development (ICT4D) with a view to generating specific applications and solutions to Africa's problems and needs. In the process, researchers will be exposed to inter-African knowledge networks and to global cutting edge research.
- Enhance the value of doctoral and post-doctoral research and in the process build up a burgeoning body of research with a direct application to STI4D and ICT4D across Africa.
- Create the opportunity for research to formulate collaborative research projects.
- Lay the foundation for individual and combined intermediate research output in the form of journal articles.
- Create enduring networks in the area of STI4D and ICT4D across universities and research institutes in Africa.

Over time, we expect AGARIC to be cumulative in its effects on knowledge networks related to the research and innovation culture combining STI4D and ICT4D and thus grow its own reproductive capacity.

Activities

The activities of each of the AGARIC sessions will include the following:

- The work the PhD scholars undertake through the presentations and papers will be evaluated and the best two papers will be eligible for a prize both to promote quality research and knowledge production and to create a culture of research excellence.
- Peer reviewed work from the PhD training seminars will also be fed into the African Journal of Science, Technology and Innovation for Development (AJSTID), whose first year has published three issues in 2009 and the International Journal of Computing and ICT Research (IJCIR)
- Annual workshop on relevant topics in science, technology and innovation will be combined to take place after the AGARICS PhD research training for new scholars. The workshop can be for 2 days after the 10 days PhD scholarly training is completed with the possibility of attendance by some of the trainees.
- A series of open lectures focusing on the various aspects of STI4D and ICT4D for the development of African systems of innovation.
- All PhD scholars in the AGARIC will be expected to give a "work in progress" report of their work and should expect to get very thorough feedbacks from their peers and the senior research scholars both formally and informally throughout an intensive 10 days mentoring work as they are also expected to do to their peers and engage with senior scholars.
- All senior researchers will provide a number of lectures on areas ranging from the latest theories and empirical research in the specific areas of training. These senior researchers will be selected from a proven evaluation of the accomplished work they have done in the field of selected training at a particular time.

Expected Outcomes

- Production of PhD research and innovation of high quality, benefiting from exposure and learning with peers and accomplished researchers in the respective fields of training
- Development of highly qualified personnel who will be able to give research leadership in universities and public and private

research Institutes and Centres across the continent

- Promotion of the progressive substitution of indigenous for imported expertise.
- Stemming the current brain drain out of Africa and provide institutional incentives for its reversal

African Masters and Doctoral Students in American Universties

To train African students in research, the authors visited Makerere University in 2008 and for one week discussed the research topics of PhD Students. In addition, through a collaborative effort, three students from Makerere university joined PhD program at Louisiana Tech university starting 2009-2010 academic session. After completing their PhD, these students are expected to go back to Uganda to impart training to other students in Uganda. With collaboration with the associate dean of graduate studies, Professor Idris Rai, the authors plan to have a stream of PhD students from Uganda at the rate of about one to two students per year to join the PhD program at Louisiana Tech.

In addition, a Kenyan student from the University of Nairobi will soon join the University of Tennessee-Chattanooga for a Masters degree in Computer Science. The authors are planning similar programs from other African universities.

Publications

Between the three of us, several publications that include a book, several conference and journals have been published and two journals launched since 2005 when all these activities started. The two journnals, which are aimed at boosting African research publications in ICT and STI are outlined below:

- *International Journal of Computing and ICT Research (IJCIR)* (Joseph M. Kizza, Editor-in-Chief). The International Journal of Computing and ICT (IJCIR) is a peer reviewed International journal with an objective of providing a medium for academics to publish original cutting edge research in the field of computing and ICT. IJCIR publishes two issues per year: June and December of each year. The IJCIR is an independent biannual publication of Makerere University, Uganda. The journal's focus is to highlight new research developments, experiences and best practices in the use of information and communications technology (ICT) across Africa. The Journal publishes papers in computer science, computer engineering, software engineering, information systems, data communications and computer networks, ICT for sustainable development, and other related areas.

- *The African Journal of Science, Technology, Innovation and Development (AJSTID)* – (Mammo Muchie, Editor-in-Chief). AJSTID is a multi-disciplinary and refereed international journal with a special focus on science, technology, and innovation in developing economies, with a special reference to Africa. It has been established on the basis of the recognized role of innovation in the development of economies and on the relative absence of research in the area, particularly in the case of Africa. AJSTID seeks to encourage research along three broad streams. This first is the role of science, technology and innovation in the process of industrial growth and development. The second stream concerns the broader area of socio-economic development. The third invites work exploring the inclusion of innovation and knowledge in cross border integration processes particularly in Africa. AJSTID solicits submissions on both these research streams at various levels: firms, sectors/ industries/ clusters, regions and countries.

Other Initiatives

Other initiatives taken by Joeph Kizza include the textbook project through which, the author has collected and is shipping books to three universities in Uganda: Mbarara University of Science and Technology (MUST), Makerere University and Muteese I Royal University (MRU).

3. Benefits to the SIGCSE community

Annually, thousands of American students visit African universities for either summer research, exchange programs or study abroad programs. They come back with rich experiences that is beneficial not only to them but also to their colleagues and their home institutions. For those who participate, exposure on issues discussed in this paper gives them a glimpse into a continent. Also several faculty members visit African academic institutions either via a Fulbright program, as was the case with one of the authors, or through other private arrangements. This kind of information is good for faculty intending to visit the continent. But the most important benefit for members of the SIGCSE is the awareness this paper brings to the community and the realization of the academic problems colleagues in Africa face on a daily basis.

4. Conclusions

There are three main activities that need to come together in order to enhance research capacity building and training, and knowledge creation to stimulate and enhance African science, technology and innovation system building. The first is establishing academic research networking and training that connects inter African research communities with each other and with the best international research networks globally. The second is to identify cognate fields from social and technical sciences with research programmes and research themes to create, acquire, build and use a body of knowledge for both the enhancement of research and quality training. The third is the organisation of a number of books, journals and other publication that will be channels of scholars research ideas and also to provide immediate outlet in Africa which may within a few years attain international rating to serve as a publishing node in addition to others to disseminate effectively analytic work, research results and newly acquired knowledge and also facilitate mechanisms for policy learning and knowledge-dialogue and interchange amongst all the relevant stakeholders.

More specifically the impact of these efforts can be summarised as follows:

- Contributing substantially to achieving the threshold of research and innovation necessary to enhance the productive power of Africa in health, wealth, food, and energy and environment security.
- Contributing to reversing the erosion of the capacity of African universities to produce new generations of quality and top academics and researchers.
- Opening of the African research and knowledge infrastructure for mobility of research ideas, researchers, training themes and other relevant cutting edge knowledge
- Enabling the African researchers and innovators to break out of their isolation by exposing them both to other African knowledge sites around the world.
- Systematically building knowledge networks between disciplines, between Africans, and Africans and researchers from the rest of the world.

Notes

1 See: "Policy Discourse vs. "Private" Opinions: The Urgent Need for African Higher Education Think Tanks", *International Network for Higher Education in Africa,* Editorial Series. Available at: http://www.bc.edu/research/cihe/inhea/editorial.html

2 Ibid.

3 *The Chronicle of Higher Education*, 25 November 2008.

4 See: Pouris & Pouris, The State of Science and Technology in Africa (2000-2004), *Scientometric,* 79, 2009.

5 Ibid.

6 Ibid.

7 Net immigration from SSA was 0.57 million in 1995 which fell to 0.29 million in 2000. It then rose to 0.7 million in 2005. An estimated one third of these were university graduates… Remittances are US 22 billion dollars for SSA (See World Bank, Accelerating Catch Up: Tertiary Education for Growth in Sub-Saharan Africa, xxxiv, 2009).

8 Adapted from SARUA's (Southern Africa Regional Universities Association): Strategic Plan 2007 – 2012 itself identifies some of the critical weaknesses that characterise the HE education system in the region.

9 Ibid.

10 See *The Chronicle of Higher Education*, 25 November 2008.

11 Berit Olson, Former Director of SAREC

12 Mammo Muchie, Presentation on Research Universities for Eradication of Poverty, UNESCO's Global Research Seminar, November 27-29, 2008.

13 DST, South African Ten Year Innovation Plan, Pretoria, South Africa, 2008.

14 See: www.nesglobal.org

References

Waema, T. M. (2009a), 'E-local governance: a case study of financial management system implementation in two municipal councils in Kenya', in *International Journal of Electronic Governance*, 2(1): 55-73.

Waema, T. M., Mitullah, W. and Adera. E. (2009b), 'Research in African e-Local Governance: Outcome Assessment Research Framework', in *African Journal of Science, Technology, Innovation and Development*, 1(1): 220-248.

Chapter 4

A Nanostep Towards Building a Sustainable Interdisciplinary Research and Educational Program between the US and South Africa

Myron N.V. Williams, Paul M. Brown, Ishrat M. Khan, Alfred Msezane, James Reed, Laurisa London & Malik Maaza

1. Introduction

Clark Atlanta University's Center for Functional Nanoscale Materials is developing transatlantic and interdisciplinary nanoscience research and education collaborations with South African Institutions. As part of this developing relationship, the Center seeks to become a major international site for to nanoscience research and education. Towards this end the Center has conducted a pilot International Research Experience for Undergraduates (IREU) Program, in which US students from a variety of Universities work alongside the Center's African Collaborators.

The University's mission and history make it especially well-suited for this undertaking. Clark Atlanta University (CAU) resulted from the 1988 consolidation of Atlanta University and Clark College, two US Historically Black Colleges and Universities (HBCUs) founded in the 1860s. Located in Atlanta GA, CAU is a comprehensive and coeducational university, classified by the Carnegie Foundation as Research University - High Research Activity.

Because of its history and its established relationships with many African institutions, the University has initiated the development of mutually beneficial research and educational partnerships in the area of nanoscience with African institutions. This history goes as far back as the early 20th century, when W. E. B. DuBois, co-organizer of the Pan African Congresses between 1919 and 1945, was also a professor at Atlanta University from 1897-1910 and 1932-1944 (Bacote, 1969). Because of this history and because the majority of its students are of African descent, this interest in the African continent continues.

The current University leadership understands that active participation in science and technology is essential to shaping a population capable of full participation in democratic citizenship and economic development in the modern world. Towards this end, international scientific partnerships not only present new opportunities, but are necessary to create the global scientific culture to improve the overall condition of humanity.

With support from the US National Science Foundation, CAU established the Center for Functional Nanoscale Materials (CFNM) in 2006 for the purposes of: 1. Conducting beneficial and innovative research for the benefit of the Nation and all humanity; 2. Increasing the number of students pursuing graduate and undergraduate degrees in the natural and physical sciences; 3. Enhancing the research productivity of its researchers.

CFNM was established at CAU to be both interdisciplinary and interinstitutional. Faculty investigators at CAU are based in the departments of Physics, Chemistry and Biological Science. Domestic partners include Cornell University, the University of Illinois, Emory University, the University of Texas Pan American, and Georgia Perimeter College. Its international partners include iThemba Laboratories for Accelerator-Based Science and the African Laser Center.

2. Nanoscience and Interdisciplinarity

Nanoscience research greatly benefits from interdisciplinary and inter-laboratory collaborations. Nanoscience deals with materials having at least one dimension in the 1-100 nm scale. It has been observed that as materials are reduced to nanoscale dimensions, often their physical properties change when compared to macroscopic systems. While scientists have been able to observe nanoscale devices in nature for decades (especially as components of biological systems), it is only in recent years that methods for synthesis and manipulation of such structures have become possible. Scientists have artificially subdivided the study of the natural and physical world into the traditional scientific disciplines, which have allowed us to reduce and understand many of the fundamental laws governing the behavior of matter and systems; however, natural phenomena themselves are not subject to such divisions. Manipulation of nanoscale materials requires use of paradigms from the biological, chemical, physical, engineering and mathematical disciplines, and is the natural sphere for interdisciplinary thinking and the training of the next generation of problem solvers. While the case for interdisciplinary collaborations has been well established as the way forward in 21st Century science, it is often difficult to create successful collaborations among individuals whose training background are very different. Similar terms may have very different meanings among different fields e.g. "transformation" means something very different in cancer biology, geometry, synthetic chemistry, physics, statistics or molecular genetics. Anthropologists studying how the mutually productive relations of knowledge (i.e. scientific exchange) take place (Rabinow and Bennet, 2007: p 4) suggest that:

> Given the emergent character of innovations in [interdisciplinary science], the precise form of collaboration called for cannot be settled in advance. However, such collaboration will involve intensive and ongoing reflection with selected … PI's on the emergent ethical, ontological, and governance problem-spaces within which their work is situated and develops.

Thus, successful nanoscience collaborations require intentional cultivation of interactions and relationships.

3. The Need for International Scientific Partnerships

In recent decades, the distribution of scientific knowledge and the pursuit of new knowledge has shifted from being largely centralized within the so-called developed countries (the G7: Canada, France, Germany, Italy, Japan, United Kingdom, and United States) and is becoming distributed among an increasing number of middle-income countries, e.g., India, Turkey, China, Mexico, Russia, Brazil, South Africa (Paraje *et al.*, 2005). The mobility of highly skilled and talented workers and the ready movement of scientific and technological information through electronic communication facilitate the redistribution of scientific knowledge at the global level. Increasingly the global science and technology industries are decentralizing not only their production facilities, but also their central research and development facilities (Niosi, 1999). This trend is expected to keep growing over the next few decades. The scientific enterprise is already global in nature and scientific educational and research institutions must be more deliberate in designing and developing programs to ensure that many more new generation scientists are participants or benefit from the globalization of science. NSF Director Arden Bement has articulated the necessity for international partnerships (NSF Press Release, 10 September 2009): "Developing effective ways to transcend traditional boundaries, and bring very different scientific cultures together for the benefit of science and society, without compromising excellence, is a critically important challenge for the Foundation".

US students studying science and technology will find that their career positions thrust them into an environment which values global diversity. Because of these transformations in the scientific and technological industries, employers will seek employees who have the ability to thrive in the new global science and technology environment. The scientific and technological disciplines in US academia must therefore focus on developing students who will be successful within this emerging global community. Historically, study abroad for American students has been largely limited to the areas of languages, arts & social sciences, with the natural and physical sciences having few international programs. In order to develop a generation of scientists who are globally engaged and comfortable working in the new environment, US institutions must develop and sustain mutually beneficial scientific partnerships with academic institutions, as well as governmental and industrial laboratories outside the US. Furthermore, as US institutions develop international scientific partnerships and exchange, they must consider countries that have historically been considered non-scientific or non-technological because it is the integration of cutting edge science in local problem-solving that will create a better society. Since English is the *lingua franca* of science, it is appropriate that these partnerships be first established with English-speaking countries, but ultimately should include nations whose primary language is not English.

The practice of science has been and continues to be global in nature as evidenced by the national origin of scientific publications and patents (Paraje *et al.*, 2005). The fact that a larger number of countries and regions are becoming scientifically and technologically accomplished translates into an interest in scientific problems which are more local in nature. This has been a driving force for the diversification of science and technology into application areas which have been ignored previously; this is particularly important for socioeconomic development in many African countries. At the same time, doing science and developing technology are becoming increasingly complex and expensive. Therefore, global and regional partnerships are crucial with respect to both the economic use of resources and maximizing access to talent and expertise. The modern international scientific community by its nature is a network of interacting laboratories through its use of the published scientific literature. A variety of new modes of electronic communication have greatly advanced collaborative possibilities, but these modes of communication may not effectively substitute for meeting and working face-to-face, particularly in collaboration across cultural and disciplinary boundaries. A World Bank-commissioned report on Building Capacity in Developing Countries states: "personal interaction helps to build trust and confidence, which are critical to forming collaborations and making them successful. Personal interaction also is key to many capacity-building activities that require the learning of physical skills and exchange of tacit knowledge (Wagner et al., 2001 p xv)".

Thus, given both the necessity and benefits of scientific partnerships, the CFNM seeks to develop international partnerships, which will strengthen its research programs and broaden the experiences of its students. International experience will not only contribute to the training of a new generation of globally-engaged scientists but will also create a cadre of young scientists who will successfully address technical and societal problems that cannot be solved locally. An example of a highly successful regional and international partnership is the International Rice Research Institute (IRRI). IRRI's mission and accomplishments are succinctly stated in the following paragraph from its website (IRRI, 2009):

> Established in 1960, we are the largest non-profit agricultural research center in Asia, with headquarters in the Philippines and offices in 14 nations. Supported by donors and partners around the globe, we are known as the home of the Green Revolution in Asia. We help feed almost half the world's population. Our mission is to reduce poverty and hunger, improve the health of rice farmers and consumers, and ensure that rice production is environmentally sustainable.

The take-home lesson of IRRI's success is that scientific breakthroughs that transform society are possible when groups of countries and institutions work together in an integrated fashion.

4. The Partnership between Clark Atlanta University and iThemba LABS

As part of the broader programs of the CFNM, CAU is developing international nanoscience research and education programs. The University's purpose includes preserving and disseminating the heritage of peoples of African descent. The great majority of the student body at CAU is of African descent, and many students at CAU and other HBCUs have a special interest in Africa. Correspondingly, part of the mission of the International Council for Science Regional Office for Africa (ICSU ROA) is to accelerate the capacity for science and technology in Africa by engaging the African Diaspora, and harnessing their skills and knowledge in Science and Technology (S&T). NANOafnet is a continental initiative that ICSU created to launch the African S&T community into the emerging multidisciplinary fields of nanoscience and nanotechnology. As part of this initiative, CAU has been designated as the North American Node of NANOafnet. One member of the Center, eminent physicist Dr. Alfred Msezane, is a native of South Africa and has been involved in international and Pan African science education and research for decades. In 2007, the Center also initiated collaborative international research and educational programs with scientists at iThemba Laboratory for Accelerator-Based Science (iThemba LABS), which is the headquarters for NANOafnet, and with the African Laser Center (ALC) in South Africa.

iThemba LABS is a multi-disciplinary research laboratory administered by the South African National Research Foundation (NRF). Based at two sites, in Western Cape and Gauteng, iThemba LABS provide facilities for pure and applied sub-atomic science using particle beams. Worldwide, these basic sciences and their applications to technology are growing rapidly. In order to rectify the shortage of the skills and human resources required to satisfy these growing market requirements and thereby build economies in which there are sufficient resources to eradicate poverty, iThemba LABS conducts training programs for South African and other African scientists. The ALC, a virtual centre of excellence, was established in 2002 to stimulate innovation, research and technology development in lasers and the application of lasers in Africa, and to provide a platform through which Africa can pool its resources to become globally competitive. In South Africa, the NRF National Laser Centre is the base of operations for the ALC. The CFNM now partners with iThemba LABS and the ALC to share material and personnel resources for original investigation and education in nanofunctional materials. Thus, the goals of the CFNM have been expanded to include a fourth purpose: *The development of a diverse globally engaged science workforce*.

Working within the traditional academic units of CAU, the CFNM is carrying out activities which will not only increase/improve research excellence but is intentional in its design to develop the next generation of globally engaged scientists. The programmatic elements that are being developed will permit faculty, students and staff from different countries and nations to work together side-by-side and in scientific teams. The Center's goal is consistent with Clark Atlanta University's focus and its understanding of the necessity and value of exposing students to international experiences. CAU's Study Abroad Program Web site states[1]:

> We believe that real education begins with the student's introduction into the global community, where he/she has the opportunity to exchange ideas and discuss differences, in their proper context, in an effort to come to a better understanding of, and have a greater appreciation for, the way of life of other peoples. This type of interaction engenders a new corpus of knowledge which did not previously reside in either of the preexisting cultures, American or the foreign one. This hybrid culture is the new source from which the study abroad student is to drink.

Further, African American students are disproportionately under-represented in the science & technology industries as well as in the pipeline for professional scientific training. We believe that an IREU conducted from CAU will help to attract and retain more African American students, among others, in scientific careers. We also anticipate an "Obama Effect", where African American students may respond to personal interaction and identification with Africa, with improved academic performance (Aronson *et al.,* 2009), enhanced confidence and commitment to advanced education and service. We have already observed some of these effects in our first cohort of students. Other organizations such as the Journey for Change Foundation promote the life-changing effects of global service experiences, particularly for young African Americans in Africa (Compton-Rock 2010). There are many other anecdotal benefits of early involvement of African American students with Africa. As scientists, we anticipate finding and recording quantifiable outcomes of our program in an academic context.

For the students and investigators from African countries, we hope that their interaction with American students will also broaden their perspective and appreciation of the international value of their own skill sets. Interaction with American students will also provide contacts that may blossom into scientific exchange or long-term collaborations with the members of the next generation of American scientists. Much as the Chinese Diaspora has played vital roles in high technology development in modern China (Smart and Hsu, 2004), it is hoped that the relationships developed between African American science students and their African counterparts will accrue long term benefits to African education and economic development. The National Science Board of the United States has stated that International S&E partnerships are essential to advancing "transformational diplomacy," because they can lay the groundwork for achieving these goals by creating apolitical connections among people, which build trust and facilitate communication (NSB, 2008). They argue that the US should use this "transformational diplomacy" to help foreign citizens better their lives, build their nations, and transform their futures, thereby enhancing, among other things, US National Security. All participants, faculty and students from the US, South Africa, and other countries, will benefit from an expanded understanding of the diverse educational systems and scientific cultures represented by the national and disciplinary origins of the collaborators.

Pilot Program: CFNM and iThemba LABS, 2009-2010

The Center for Functional Nanoscale Materials applied for and received an NSF CREST supplement in 2008, to carry out a pilot program to develop meaningful and mutually beneficial collaborative relationships with the African Laser Center, Pretoria, and iThemba LABS, Somerset West, South Africa. Investigators from Chemistry and Physics departments at CAU have undertaken a research collaboration developed around the research topic "Gold Nanoparticle/Synthetic Functional Polymer Nanofibers (Nanocomposites) by Electrospinning: Novel Materials for Amperometric and Photonic Bionsesors". In keeping with our educational philosophy of maintaining the inexorable link between research and student training, the project was initially driven by a pilot International Research Experience for Undergraduates (IREU) called the Nanoscholars Abroad Program. The collaborative study was initiated with the intention that it would lead to the development of mutually beneficial, long-term research and education collaborations among CFNM, ALC and iThemba LABS. A further goal was to move towards development of a sustainable and externally funded IREU with Clark Atlanta University established as a North American IREU site for South Africa. The research team is composed of senior scientists, undergraduate and graduate students trained in biology, chemistry, mathematics and/or physics. The current senior scientists for this collaborative effort are: Drs. Khan, Msezane, Reed, Williams (CAU, CFNM), Maaza (iThemba and NANOafnet) and Aboubaker Beye (ALC and NANOafnet). The NSF supplement allowed us to pilot an IREU program during the June-July of 2009, in which four (4) summer IREU students engaged in theoretical and laboratory nanoscience research for eight weeks divided between the Atlanta and South African sites. For this pilot effort, we developed a "student-centered" program sensitive to the needs of not only CAU students, but also to the participants in CAU summer research program, many of whom come from regional HBCU's with less research capacity.

Program Requirements

To advertise the Nanoscholars Abroad Program, we mailed more than one hundred posters promoting the program to US institutions, and every effort was made to ensure all the HBCUs and other Minority Serving Institutions (MSIs) were informed about the program. The CFNM Web site was redesigned to include a detailed description of the program. Participation in the program was open to undergraduate students majoring in one of the science or mathematics disciplines, who planned to attend graduate school after graduation, and were US citizens or permanent residents. Applicants were required to be rising sophomores, juniors or seniors in good academic standing at their home institutions. Interested students could apply as early as possible by completing the application found on the CFNM Web site,[2] which they could print out and mail in, or return via e-mail. All applicants had to have a passport that was valid throughout the program period, or immediately apply for a passport. Applicants were required to submit an official undergraduate transcript as well as an essay (200 words or less) on their science interests and what they expected to accomplish in their career. In addition, two letters of recommendation were required, at least one of which had to be from one of the candidate's college science instructors. The deadline for applying was April 1, 2009. Completed applications received by April 1, 2009 were quickly processed and by April 15, 2009, we had informed successful applicants of their acceptance into the program. Accepted applicants were given a timeline, to confirm to us that they had had the necessary vaccinations, passport requirements, insurance and other trip preparations.

Applicant Selection

For the Nanoscholars Abroad Program, a total of twenty-nine complete applications were received from students at 14 undergraduate institutions (including 11 MSIs). Of the total applicants 75% were female, 29% were second year, 18% third year, 41% fourth year, and of those who had declared majors, 50% were chemistry, 29% were biology, and 21% were mathematics. Cumulative grade point averages (GPA's) of applicants ranged from 2.5 to 3.89, with a mean of 3.29 and a median of 3.5. The applicants were not required to indicate race or ethnic origin in their application. We did seek to achieve gender balance and to ensure that different types of institutions and majors were represented. From the pool of applicants, the committee (composed of Drs. Khan, Williams and Msezane) selected three (3) females and one (1) male from four (4) different institutions (CAU, Savannah State University, Winston-Salem State University, and Emory University). Three of the accepted applicants were fourth year and one was a third year student, with majors in Chemistry and Math. Three of these students had previously travelled abroad (Mexico, Germany, and Ghana); all had previously participated in undergraduate research at their home institution and/or other institutions. The group was multiracial, with students self-identifying as Black and White. Applicants were given ten days to respond to the offer.

Program Implementation

The summer 2009 (June- July) program for the four (4) summer IREU students took place over eight weeks, two weeks at CFNM and six weeks at iThemba. During the first two weeks, the students were housed in the Clark Atlanta dormitories, and board was provided in the student cafeteria. The IREU students were joined in the dormitories and in many scientific activities by eight (8) undergraduates in the Center's domestic summer REU program, who were engaged in research for 6 weeks at CAU. Many of the students, prior to being accepted into the program, had only limited exposure to the nanosciences. At this time they were introduced to nanoscience research, which included hands-on fabrication of nanofibers by electrospinning, and fiber characterization by Scanning Electron Microscopy (SEM) and Atomic Force Microscopy (AFM). During this period, all four IREU students were integrated into the collaborative research team project and specific roles/functions were assigned. Ms. Laurisa London, a doctoral student at Clark Atlanta University, served as Nanoscholars Abroad peer leader. In addition to hands-on experiments, all students participated in a Problem Based-Learning workshop and a weekly Journal Club led by CAU graduate students or faculty. Papers discussed included a transcript of Richard Feynman's 1959 seminar at the American Physical Society meeting (introducing the idea of nanotechnology), an American Scientist review on Biological Machines, a commentary on high-tech research in the developing world, as well as primary journal articles presenting nanotechnology experiments in various contexts (See: *Journal Club Papers* in References). These papers provided a forum for lively discussion about the nature of the nanoscale, organization of the research enterprise, and the logic, method and reporting of discovery science.

As part of preparing the students for living abroad, a booklet with health information, the travel checklist, cultural and practical advice on daily living abroad in South Africa and program rules and guidelines was developed and made mandatory reading for the students. The Clark Atlanta University study abroad Director, Dr. Paul Brown, held a workshop for the Nanoscholars Abroad on traveling and living abroad. In this pilot program, the faculty explicitly avoided discussions centered on history, race and ethnicity in the US or South Africa, in order to avoid contaminating this generation of students with Apartheid-cognizant experiences of a previous generation. The students were given the opportunity to view the film "21 Up South Africa: Mandela's Children" (Gibson, 2007). This Angus Gibson documentary follows the lives of 11 South African children from various ethnic, regional and economic backgrounds, from age 7 to 21. In future iterations of the IREU, we anticipate adding a workshop from a scholar well versed in the dynamics of social change in Southern Africa. In Cape Town, the team was received by Dr. Malik Maaza, Chairman of NANOAFNET at iThmeba LABS, who arranged housing for the American team members in condominiums with excellent views of the Indian Ocean in the town of Strand. Strand (Afrikaans for 'beach') is a seaside resort town on the northeastern edge of False Bay and at the foot of the Hottentots Holland Mountains, about twenty-five kilometers from iThemba LABS. Local transportation was arranged by Dr Maaza.

The six-week stay in South Africa was productive for the collaboration. In addition to the students carrying out quality research in nanoscience, they had the opportunity to interact with physicists, chemists, engineers and computational scientists from throughout Africa (Morocco, Cameroon, Nigeria, Tanzania, as well as South Africa) and other parts of the world. Faculty participants were able to further develop the collaboration with our South African colleagues, and generate new collaborative ideas. The students, in addition to working at iThemba LABS, also carried out experimental work at the Stellenbosch University (atomic force microscopy [AFM], and preparation of electronic chip by photolithography), University of Cape Town (transmission electron microscopy; TEM) and University of the Western Cape (Fourier-transform infrared spectroscopy; FTIR) and to visit the Tshwane University of Technology (Pretoria). These institutions exposed students and faculty to a substantial part of the diversity of higher education in South Africa; the University of Cape Town and Stellenbosch were historically white and the flagship institutions for scientific research. The University of Western Cape and Tshwane University of Technology were derived from Historically Black Universities (HBUs) that served "Colored and Black" students during Apartheid. Additionally, Dr. Maaza arranged for visits by participants to the National Research Foundation and National Laser Center, in Pretoria. During the six-week stay in South Africa, at any given time, our programmatic structure included the presence of at least one CFNM/CAU faculty with the student team. In fact, except for a one-week period, at least two CAU faculty members were present with the student team. This was an important aspect of our program and most likely an important factor in the smooth operation of the program. Drs. Msezane and Khan were replaced by Drs. Reed and Williams at the halfway point of the program. The faculty not only worked as part of the team but also facilitated several trips for the students during the weekends.

The Collaborative Research Project

Essential to the success of the program was an ongoing research collaboration between the laboratories in the CFNM and those at the iThemba LABS. The collaboration was concerned with creation of functional nanostructures that have had a wide variety of applications including their utilization as the active components in nanobiosensors. A number of investigators have developed immunosensors based on localized surface plasmon resonance (LSPR) of metallic nanostructures (Erickson *et al.*, 2008). With the use of simple optical devices, LSRP of nanostructures is a strong candidate for the development of miniaturized, highly sensitive, and high-throughput devices for the detection and monitoring of biomarkers. Low cost, point-of-care diagnostic methods are important for monitoring protein and nucleic acid biomarkers associated with exposure to and progression of a variety of diseases prevalent in Africa (e.g. tuberculosis and HIV), and corresponding health care. We propose that gold nanoparticles can be arrayed for creation of multiplex sensors by embedding in polymer nanofibers prepared by electrospinning. The importance of gold as a "local" natural resource in South African was also an element in project design. The expertise of various members of the collaboration makes this project highly feasible.

The students learned to prepare the gold nanoparticles for the 2009 IREU investigation by two methods. The first is a technique known as the liquid-liquid interface method (Premkumar *et al.*, 2007). The nanoparticles were formed at the interface of a solution of a gold derivative dissolved in toluene, and an aqueous layer containing a reducing agent. The second method used to form gold nanoparticles was a one-phase synthesis technique (Rao *et al.*, 2003). The particles were prepared in a solution containing a nonionic surfactant and a gold salt. In one-phase synthesis, some of the gold nanoparticles that formed were suspended freely in solution, but most settled to the bottom of the reaction vessel. Gold nanoparticles prepared by the liquid-liquid interfacial method show an even spacing among the nanoparticles, from which it can be inferred that there is a bonding interaction occurring among the nanoparticles. The gold nanoparticles ranged in diameter from approximately 3 to 43 nm. Furthermore, some of the gold nanoparticles formed with faces that were perfect polygons. Investigation of gold nanoparticle growth dynamics was carried out using UV-Visible spectroscopy and measurements of the refractive index. Nanofibers were electrospun from a solution containing poly(vinylalcohol) and gold particles. Prior training of the US graduate and undergraduate students facilitated this aspect of the project. Careful control of the viscosity of solutions used for electrospinning are necessary, in order to form long fibers with small diameters. A 21% (w/w) solution of polyvinyl alcohol was combined with gold nanoparticles from the 0.05 mM one-phase synthesis. This nanocomposite solution was electrospun onto several substrates: glass, indium tin oxide (ITO), silicon, ITO on plastic and aluminum foil. Collaborating engineers at Stellenbosch University taught the students lithographic techniques for creating the kind of patterned surfaces that ultimately would be used in sensor devices. Characterization of the fibers and gold nanoparticles was carried out using AFM, SEM, TEM, ATR, UV- Visible spectroscopy, and STM. Scanning electron microscopy was used to examine the gold/PVA nanofiber composites that had been electrospun onto several different substrates for different amounts of time. These studies indicate that the electrospinning was successful; however, there was very little alignment of the fibers. The ITO-PET and the aluminum substrates showed almost no alignment of the fibers, and the gold/silicon lithography and the plain silicon substrates showed some alignment towards the edges of the substrates. Fiber alignment will be critical for development of multiplexed sensors. The fibers also appeared to have a larger diameter than desired (<100nm). We anticipate that the diameter can be decreased by increasing the voltage used for electrospinning. Currently, we are developing conditions for preparing fibers with smaller diameters and we are also developing strategies to ensure a better distribution of the nanoparticles within the electrospun fibers. Once the appropriate conditions have been developed, functionalized nanofibers will be prepared and then used as the active component in nanobiosenors.

The mathematics student and theoretical physicist mentors engaged in calculations using Web accessible databases and software (gnuplot) to investigate low-energy electron scattering from ground and excited state gold and platinum atoms. The purpose was to understand the structure and the dynamics of electron attachment at the atomic scale, resulting in anion formation as resonances. The collaboration, that also involves the Queen's University of Belfast, UK has now obtained results from studies using Regge-pole methodology (Felfli *et al.,* 2010a; Felfli *et al.,* 2010b). The calculations used the US DOE Office of Science Supercomputing facilities and those of the Queen's University of Belfast. The resulting two papers have now provided the novel fundamental mechanism for nanoscale catalysis of hydrogen peroxide using gold and palladium nonoparticles as catalysts. The research is ground breaking in that it provides for the first time ever the fundamental atomic mechanism to explain the observed exceptional catalytic properties of Au nanoparticles and why the combination Au-Pd possesses even a higher catalytic activity than Au or Pd separately when catalyzing H_2O_2, consistent with recent measurements. On July 31, 2009, the CFNM team returned to Atlanta after a highly successful summer research engagement.

Program Evaluation

The 2009 pilot IREU has been evaluated and ongoing evaluation is still in progress. Our evaluator for the program was Dr. Willie Pearson, Jr., Professor of Sociology, School of History, Technology and Society at Georgia Institute of Technology. Questionnaires to record the students' opinions about their knowledge, plans, and other aspects of the program were administered at the beginning of the program (pre) to both IREU and domestic REU students, and again in a more extensive form at the end (post). Several of the applicants we accepted into the pilot IREU Program had also applied to and been accepted in other summer programs. Their declared reasons for choosing participating in this program included:

- "Experience of traveling abroad, cultural diversity in the science world"
- "Stipend, faculty, length of program, University reputation"

The most important value of the Nanoscholar Abroad Program for the students included:

- "Being able to work with a wide range of students from different places, therefore being able to see different viewpoints and different ways of attacking a problem"
- "I found that I much rather work alone with the help of faculty"
- "An immersive experience of cultures, personal experiences and scientific research"
- "Domestic and overseas research experiences would provide a different value for me because [my overseas experience allowed] me to gain another perspective in [how one's] life experiences effect how one sees research"
- "It was a very peaceful and learning-friendly environment"
- "This program provided a great learning experience in the field of science, and learning about other cultures"

The program had positive effects on the future career plans of our students:

- "I'm very interested in working as part of the global economy"
- "The global influence, not only where we are interacting with local South Africans, but also individuals from all over the world"
- "I have declared my interest in a career in research and teaching"
- "[I'm now] interested in doing research in nanotechnology in graduate school!"
- "I would like to continue work on the research I have started with the program

As illustrated in Figure 1, the student group's perceived knowledge of the concepts underlying nanotechnology increased by a more than a full point (on a 5 point scale). Similarly, their interest in pursuing a career in nanotechnology significantly increased, and there was a marginal increase in the students' desires to pursue a career in scientific research. Their overall interest in pursuing graduate study did not significantly change, nor did their excitement about how nanotechnology can improve the world, nor their interest in nanotechnology research. In comparison to the group of domestic REU students, the Nanoscholars Abroad began with greater interest in employment in a global industry. Strikingly, this interest of IREU students increased following their research experience in South Africa, while the interest of the Domestic Nanoscholars (control group) in global industries did not change significantly.

Figure 1: Effect of Nanoschaloars Abroad Experience on Student Attitudes & Plans

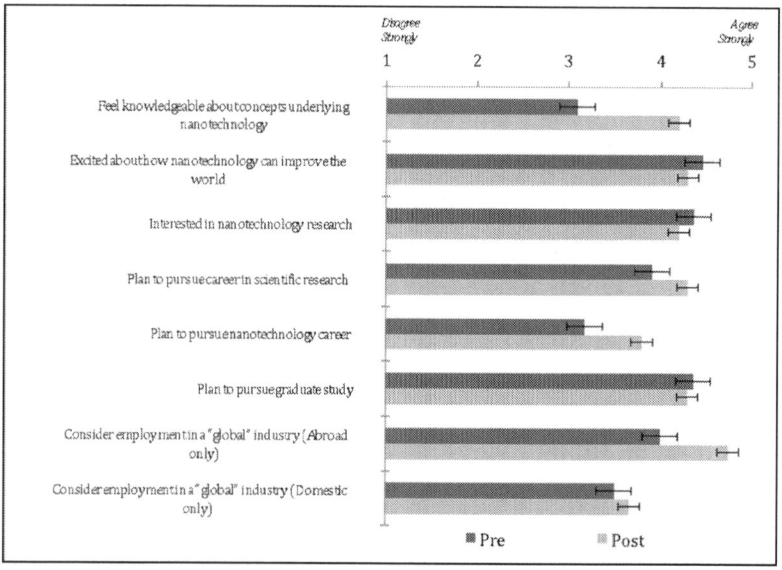

Source: Maaza M., *et al.* (2009); *Note:* Error bars represent 1 standard deviation.

A number of modifications to the program were also suggested by our experience. Most of the students worked interchangeably on the wet chemistry project, while the theoretical studies were rather too isolated. Assignment of each student to different mentors on the South African site, combined with regular group meetings involving all the students should provide both the students with greater ownership and formalized interdisciplinary interaction. The pairing of Nanoscholars with near-peer students (graduate or undergraduate) at the South Africa site should permit even more substantive interpersonal and intercultural interaction. If we are able to identify US participants earlier in the spring, some of the cultural and scientific orientation can begin via Web resources before their arrival in Atlanta. Thus, their stay at CAU could be reduced, and students would have a longer time to complete their projects at the host institution. Further, we hope to develop a mechanism by which some of the experience of the students can be disseminated upon their return to the US. They should be able to relate their experience of international travel and work to local K-12 students and teachers, and the technical skills acquired can be used at their home institutions, ideally by involving faculty advisors at the home institution in the planning and development of their projects.

It is clear that effective interdisciplinary collaboration requires special skills in interpretation and explanation of different traditions, as well as a high level of openness about the significance and reliability of theories and data. The process of learning how to collaborate is similar to that which travelers and immigrants must undergo when crossing national and continental borders (Montoya, n.d.).

5. Conclusion

We suggest that by simultaneously positioning ourselves and our trainees in a research environment that is both international and interdisciplinary, the adaptive proficiency learned in the cultural arena will infiltrate our scientific thinking. The scientific innovation of the work performed by the Nanoscholar group in the first iteration of the program demonstrated the ability of supervised undergraduate students to adapt parts of established protocols from the literature towards a new combination of goals. Aspects of the work have indeed been ground-breaking and entirely novel. The program clearly provided the students with an outstanding range of interdisciplinary experience. In addition, our interactions reminded each of the involved faculty of aspects of scientific method that are common across disciplines, made us aware of some of the current advances and challenges outside our own specialties, and helped us to shape the forward motion of a collaborative project that makes excellent use of our collective capabilities. Furthermore, US and African faculty shared challenges, strategies and ideas about how to enhance science pedagogy for undergraduate and graduate students and substantive collaborative research.

The formal collaboration and partnership of CFNM with iThemba LABS has been successfully implemented. Based on our research outcomes and a better understanding and appreciation of our colleagues and resources in South Africa, we will keep moving forward in a very holistic manner, i.e. science, culture and friendship integrated in all our efforts. This is the first time we operated a summer IREU, and it was an encouraging accomplishment, resulting in positive scientific and educational outcomes. We intend to use our programmatic element from this summer IREU to formally develop Clark Atlanta University into an international REU site for South Africa, which we believe will provide both a model for other S &T education programs and for its direct scientific impact.

Notes

1 See: http://www.cau.edu/Academics_Philosophy.aspx

2 See: www.cfnm.cau.edu

References

Aronson J., Jannone, S., McGlone, M., Johnson-Campbell, T. (2009), 'The Obama effect: An experimental test', in *Journal of Experimental Social Psychology*, 45 (4): 957-96.

Bacote, C.A. (1969), *The Story of Atlanta University*, Princeton NJ: Princeton University Press.

CAU Office of International Education/Study Abroad. See http://www.cau.edu/Academics_Office_of_International_Education_and_Study_Abroad.aspx

Center for Functional Nanoscale Materials. See: http://www.cfnm.cau.edu/.

Compton-Rock, M. (2010). "The Angel Rock Project:: Journey For

Change: Empowering Youth Through Global Service", Available at: http://www.angelrockproject.com/arp/projects/journey_for_change.asp (Accessed on: 29 April 2010).

Erickson, D., Mandal, S., Yang, A.H., Cordovez, B. (2008). 'Nanobiosensors: optofluidic, electrical and mechanical approaches to biomolecular detection at the nanoscale', in *Microfluid Nanofluidics*, 4 (1-2): 33-52.

Felfli, Z., Eure, A.R., Msezane, A.Z. and Sokolovski, D. (2010a), 'Low-energy electron elastic scattering cross sections for excited Au and Pt atoms', in *Nuclear Instruments and Methods in Physics Research Section B*, 268: 1370.

Felfli Z., Msezane A.Z. and Sokolovski D. (2010b), "Complex angular momentum analysis of low energy electron elastic scattering from lanthanide atoms", in *Phys. Rev. A*, 81: 042707

Gibson, A. (2007), "21 Up South Africa: Mandela's Children", Icarus Films, NY.

International Rice Research Institute (2009), Available at: http://beta.irri.org/index.php/Home/Welcome/Frontpage.html (Accessed on: 10 September 2009).

Maaza M., Williams M.N.V., Khan I.M., Msezane A., Reed J.L. (2009), 'The International Research and Educational Programs of Clark Atlanta University's Center for Functional Nanoscale Materials', in Susan E. Cozzens and Pablo Catalán (eds.), *Proceedings of the 2009 Atlanta Conference on Science and Innovation Policy*, IEEE. Available at: http://smartech.gatech.edu/handle/1853/32322.

Montoya, M. (n.d.), 'Beyond the Scientific Pipeline: Toward a Pluralistic Science for the 21st Century', Available from: mmontoya@uci.edu.

Niosi, J. (1999), 'The Internationalization of Industrial R&D: from Technology Transfer to the Learning Organization', in *Research Policy*, 28 (2-3): 107-117.

NSB, National Science Board (2008), *International Science and Engineering Partnerships: A Priority for U.S. Foreign Policy and Our Nation's Innovation Enterprise*, February, Arlington, VA: National Science Foundation. Available at: http://www.nsf.gov/pubs/2008/nsb084/nsb084.pdf

NSF Press Release (2009), Available at: http://www.nsf.gov/news/news_summ.jsp?cntn_id=110850 (Accessed on: 10 September 2009).

Paraje G., Sadana R., and Karam G. (2005). "Increasing International Gaps in Health-Related Publications." in *Science,* 308 (5724), 959.

Premkumar, T., Kim, D., Lee, K., and Geckeler, K.E. (2007), 'A Facile and Efficient "One-Step" Synthesis of Au° with Tunable Size', in *Gold Bulletin*, 40 (4): 321-327.

Rabinow, P. and Bennett, G. (2007), 'A Diagnostic of Equipmental Platforms', *ARC Working Paper,* No. 9. Available at: http://anthropos-lab.net/documents/wps/a-diagnostic-of-equipmental-platforms/.

Rao, C.N.R., Kulkarni, G.U., Thomas, J.P., Agrawal, V.V., Gautam, U.K. and Ghosh, M. (2003), 'Nanocrystals of metals, semiconductors and oxides: novel synthesis and applications', in *Current Science*, 85 (7): 1041-1045.

Smart, A. and Hsu Y-J. (2004). "The Chinese Diaspora, Foreign Investment and Economic Development in China." In *The Review of International Affairs*, 3 (4): 544 – 566.

Wagner, C.S., Brahmakulam, I., Jackson, B., Wong, A., and Yoda, T. (2001), *Science and Technology Collaboration: Building Capacity in Developing Countries*, MR-1357.0-WB, Santa Monica, CA: RAND.

Sample Journal Club Papers

Arunachalam, S. (2006), 'Democratizing potential of Open Access: Perspectives from the Developing World', *Paper Presented at Africa Study Centre*, 6-8 September, Leiden.

Feynman, R.P. (1960), 'There's Plenty of Room at the Bottom: An Invitation to Enter a New Field of Physics', in *Engineering and*

Science Magazine, February, 23 (5). See: http://media.wiley.com/product_data/excerpt/53/07803108/0780310853.pdf.

Goodsell, D.S. (2000), 'Biomolecules and Nanotechnology', in *American Scientist*, 88 (3): 230-237.

Ko, Y-J., Maeng, J.H., Ahn, Y., Hwang, S.Y., Cho, N-G. and Lee, S-Y. (2008), 'Microchip-based multiplex electro-immunosensing system for the detection of cancer biomarkers', in Electrophoresis, 29 (16): 3466–3476.

Mohapatra, S.K., John, S.E., Banerjee, S. and Misra, M. (2009), 'Water Photooxidation by Smooth and Ultrathin α-Fe2O3 Nanotube Arrays', in Chem. Mater., 21 (14): 3048–3055.

Suri, S.S., Fenniri, H. and Singh, B. (2007), 'Nanotechnology-based drug delivery systems' in *Journal of Occupational Medicine and Toxicology*, 2 (16). See: http://www.occup-med.com/content/pdf/1745-6673-2-16.pdf.

Watari, F. (2008), 'Reaction of cells and tissue to material nanosizing', in *Transactions of the Materials Research Society of Japan*, 33 (2): 209-214.

Acknowledgement: Portions of this manuscript have previously appeared in: Maaza M., Williams M.N.V., Khan I.M., Msezane A., Reed J.L. (2009), 'The International Research and Educational Programs of Clark Atlanta University's Center for Functional Nanoscale Materials', in Susan E. Cozzens and Pablo Catalán (eds.), *Proceedings of the 2009 Atlanta Conference on Science and Innovation Policy*, IEEE. Available at: http://smartech.gatech.edu/handle/1853/32322.

Chapter 5

The Challenges and Opportunities of Re-inventing Higher Education as Centres of Research Capacity Building

The Mammo Muchie & Angathevar Baskaran

1. Introduction

Just as the world economy is uneven, knowledge, research, higher education and innovation are embedded in different countries in the world very unevenly. In all the ratings the top ten universities often are from the USA and the UK. Very often the top university from Africa in 2010 is the University of Cape Town at 161! This shows that knowledge and research are unevenly spread calling for a knowledge redress where those at the bottom of the pyramid of the knowledge universe need to develop strategies to climb the research ladder. The issue of those who are research poor and others that are research rich needs to be addressed. There is not only a bottom of the pyramid in the economic sphere, there appears to be even an established and non-changing division of a research bottom of the pyramid where the top ten universities are often from the same countries and reflecting similar cohort of universities that continue to be ranked up and down within the same range. How can this established or fossilized division between the research rich and the research poor change?

The Globelics experiment can be seen as one small way to broach this division between the research rich and research poor by making it possible for researchers to be mobile and encourage movement from the North to the South. Globelics appears to have emerged as a research cooperation model to move the best researchers from the North to the South where the research effort requires to be enhanced. Globelics can provide, if it continues with the way it has been addressing research inequality and uneven distribution, a new model of research cooperation that can potentially influence other research frameworks like the coming European FP8. In this paper we highlight how research support must move from donorship to partnership and not use research poverty as a means to use aid resources to set research agendas for the South particularly in Africa. This is a critical change that is needed to develop a new model of research partnership away from donorship and anchored on the foundation of clearly explicated and understood partnership.

2. Changes in the Wider Context and Implications for Universities

Higher education in general and universities in particular have undergone changes due largely to the changes from the provision of higher education and research as a public good to the emergence of a variety of quasi-market systems (see Baker and May, 2002). On the one hand knowledge capital is seen to have surpassed the value of physical assets, on the other universities that are the main producers of knowledge both in intangible and tangible forms are facing a context of pressures on the way they function and how they should function. Some analysts celebrate the value of universities as key to 'learning and development processes' (Archibugi and Lundvall, 2001). Others emphasize the crises of the universities linked to the rise of managerialism and commercialization (Ziman, 1994). This broader context in which universities are valued for spearheading the knowledge economy and equally lamented for the mis-direction of their traditional role by serving academic capitalism instead has hugely brought into question mark what it is that universities are and are likely to become.

It is under this general ambivalent context of either being under the crises of universities or perhaps being under better opportunities for renewal owing to the universities' submission to the market validation of their training and research efforts that the universities in the poorer countries too have to operate and navigate a future.

It is widely known that many higher education institutions in many developing economies have been confronted with insurmountable difficulties in generating and sustaining programmes of research and training capacity and productivity including using research results in their economies and societies. Higher education is often seen as a luxury or even a 'white elephant' in poor countries. This has the unfortunate impact of de-legitimizing universities in Africa and more broadly countries that went through the process of structural adjustment. Many universities have degraded and often become glorified high schools mainly preserving their teaching functions and knowledge dissemination to the determinant of research development and knowledge creation. This has deprived countries from relying on their universities to create a system of research based- higher education and human resource development.

The key identity of a university is the degree to which a research activity is embedded with institutional autonomy, and academic freedom to raise difficult problems and fearlessly investigate and find knowledge and insight. Developing countries that control universities whilst outsourcing the funding of their activities to donors severely damage the identity of universities. Universities under state control ended up not only teaching places but also part of the civil service lacking autonomy, the hallmark and distinction of what makes them universities. Lack of autonomy and academic freedom put a shadow over the whole research effort putting in question mark the hallmark of the identity of being or becoming a university. Whilst the universities face numerous obstacles such as pressure to increase enrolment, to charge fees for students, failing to equip research facilities and libraries, they still remain the primary institutions for doing any research in developing economies. Even with alternative centres such as research NGOs on the rise, the universities remain the place for doing whatever research is undertaken in many poor/low income economies.

The challenge is thus to make the university a research university along with making the national system of innovation with mutuality and reinforcement. How to bring this about in poorer economies with barely rudimentary national systems of innovation becomes a formidable challenge.

3. The Hazard of Selling National Policy Autonomy for Cash to Universities

We suggest that the simple test of classifying countries as very poor is when a Government accepts what is called budget support or cash knowing full well that by doing so it is losing control of national policy making by surrendering such control to those who give the cash. If a country can refuse to receive cash, when it is aware that such cash can be potentially a poisoned chalice since receiving the cash or borrowing comes with the hazard of control over its national policy, it shows confidence that it may be finding ways to create a national economic and production system to deal with its conditions of indignity, inequality and poverty. India is one such economy that shows this confidence. We can say that if a country has not refused budget support only for reasons of pride, it can be reasonably assumed that its refusal signifies preference to undertake a national policy for development. However, most poor economies especially in Africa rely on budget support from the international donor aid system. Those who provide the aid often attached conditions, in some cases the conditions can turn into policy advice. An important case of policy advice that came in the 1980s from the World Bank was the so-called rate of return on education investment argument. The centrepiece of this argument is that the return on investment in universities in developing economies does not justify the outlays. The investment should rather be spent on building primary education. Even primary education should not be cost-free to the parents (World Bank, 1971, 1974a, 1980, 1988).[2]

There were both crises of education and crises of universities in many poor and unequal societies. It seems that many poor countries have not put behind them the crises of the university. If in poor and radically unequal societies, the demand on the poor to pay user fees for education is imposed, there is prima facie reason to suspect that the commitment to education and knowledge production may be shallow.

At the root of this discursive agenda driven by donors is the world economic crises that started in the 1970s.The economic crises put severe strain on the foreign exchange flow to many poor economies specially those that relied on oil imports. In its education sector reviews, the World Bank in 1974 claimed: "...educational systems in developing countries are all too often ill conceived and are not adapted to their developmental needs" (World Bank, 1974b). The World Bank argued for shifting resources to basic education from secondary and tertiary education by suggesting that higher education, though important for development, has failed in the poorer part of the developing world.

According to the World Bank, in very poor developing economies, universities and other higher education bodies produce not much original new knowledge. They produce many teachers or unemployable graduates who seem to be poorly trained. They produce fewer engineers, scientists, skilled manpower that is nevertheless critical for development. University education suffers from the quality of output while the cost of such output is claimed to be disproportionately high. The recommendation that flew from this diagnosis was simple and forward: minimize the cost of university higher education, change financing, privatize institutions and functions, demand that students pay for the cost of their education by increasing student fees or creating loan schemes. In addition, there were recommendations such as introducing cost-sharing and re-orientation of academic programmes, efficiency, increasing part-time programmes, reduction of non-academic staff by using students, increasing class size, expanding distance and extra- mural programmes.

The twin problems of lack of justifying quality of output and expensive financing of the cost of university based higher education have been used to de-legitimize the university in many poorer economies, unlike those in the industrialized world where demands to make universities market driven rather than public driven is at the heart of the emerging controversy and debate on the crises of universities.

4. Reinserting Centrally Universities in National Systems of Innovation

If the assumption that nations require knowledge to enter into international competition is correct, there can be no argument that can be used to deny nations from building their knowledge base to create wealth. ACADEMIC CAPITAL or Knowledge is now a necessary condition for making and creating the wealth of nations. The OECD recognizes knowledge as the driver of 'productivity and economic growth' (OECD, 1996: 3). The term 'knowledge-economy' has emerged based on the assumption that knowledge has become a major competitive stake for nations. Universities have remained the key sites for knowledge creation and training and their role is likely to increase rather than diminish in the 21st century. From the 1960s to 1995, the number of graduates across the world increased from 13 million to 82 million (UNESCO, 1998). With the massification of education, it appears this number must have grown massively by 2006! The European Union aims to invest 3% of GDP in R&D by 2010 to make Europe the most dynamic and competitive knowledge based economy (European Commission, 2000). Some countries like Sweden are said to have 55% of the relevant population category with university degrees. This will no doubt not only create possibly users who are literate, but also in terms of overall cultural standards, the society achieves a position that puts knowledge at the heart of social-economic progress and political debate and development.

Many developing economies especially in Africa created universities in the 1960s to develop specific skills needed for African economic construction and the civil service for the bureaucracy, and training teachers for schools. The early part stressed knowledge development that can facilitate manpower planning and human resources. The economic crises of the 1970s created dilemmas that triggered a crisis in the universities. Doubts emerged related to the poor quality of output, mismatch of graduates to labour market needs, funding challenges both due to constraints by internal and external sources. It is under these conditions that economic calculation as criteria to validate investment in higher education development came. The positive role of universities and their identity and value became compromised.

At the core of the attack on the universities is the weak linkage of universities in the context of national development. When nations cannot deploy effectively national institutions such as universities, knowledge, incentives and policies, they lack the ability to generate a functioning national system of innovation. Under this context it is often the case that the role of a key component of NSI such as universities that produce the research and knowledge system in a country can be devalued. The key to reverse this negative trend is thus to integrate universities as the premier institutions of knowledge-making and production as a core part of a functioning system of innovation. In many African countries the absence of the analytical framework for building functioning national systems of innovation is at the core of the crises of the universities that is still going on right now. The problem is not simply lack of funding; it is much deeper than that. It lies in the failure to conceptualize the central role of knowledge production and diffusion and utilization for bringing about the structural transformation of economies, societies and polities in poorer and developing economies of Africa, Latin America and Asia.

There is a need conceptually to re-centre universities for the making of the national systems of innovation (NSI); and conversely to making innovation systems that deploy universities as the key institution for knowledge production. The making of universities that can create knowledge, quality output, and quality graduates with relevance to the needs and demands of economic change, social and political change, technological change, scientific breakthroughs and life styles and behaviour changes is critical to the re-insertion and re-imagination of universities as key actors for transformation.

The social-economy wide system and the university sub-system in the context of developing economies need to be integrated to enhance quality output in research and training. The NSI and triple helix perspectives permit the linking of the universities that carry out research with the wider economy, state and society. This establishes the universities as a critical component of the economy and the society. The question of how the knowledge produced by the universities is managed to meet social goals is important in the context of the evolution of the particular NSI.

5. The GLOBELICS Experiment

GLOBELICS has evolved as a research community by interfacing innovation studies and development studies. It has been conceptualized as a network for 'researchers without borders' (Luc Soete, 2003). Its main drive is to take science and knowledge to the 'world countryside' from its complacent existence from the 'world city'. All the major Globelics annual conferences have taken place in many developing countries and are likely to continue going deeper and deeper into more remote sites where the reach of world class contemporary research is inaccessible and difficult. The sheer ability to pull off a learning dynamics in difficult circumstances is indeed an achievement in and of itself. The idea is to take the best research and researchers to areas of the world where the opportunities for knowledge interaction, research interaction and training interaction are less than those that exist in the world cities.

The model Globelics offers is vastly different from universities acting as foreign enterprises creating campuses and different types of lucrative and business arrangements. Globelics prides itself in empowering those who benefit most by it namely Southern researchers. The Southern researchers with some assistance from the international scientific board of Globelics consist the leading researchers on the economics of innovation.[3] There is a judicious combination to encourage the best flow of research with the latitude to local researchers to be free to design the research they think important and relevant in the broad field of science, technology, innovation for development. They design most of the activities of research communication, exchange and planning by raising also nearly 100% of the funding from local sources.[4] This has made Globelics to evolve as a self-organizing community of researchers capable of generating world-wide learning and networking. This has increased the opportunity of many researchers to be linked in the world network of researchers with the opportunity to co-publish with senior and younger researchers across a global pool or spectrum of research. It has given Southern researchers not only to continue links with Northern researchers but also to create new ties with Southern researchers. The result has been impressive given that the cost for the production of the result has been often absorbed locally. A number of research programmes have been established and to date the results are appearing to materialise. Many books have come out; at least one or two books each year that get launched at the Globelics meetings during the last four years. A number of Journals are appearing including AJSTID (www.ajstid.com). This trend is likely to continue as precedents have been set by previous communities in Rio, Beijing, Pretoria, and Trivindrum in India, Saratov in Russia, Mexico City, Dakar, Kuala Lumpur and in Argentina in 2011 and Shanghai in 2012. There is already possibly others bidding for the years after 2012. Interest looks to remain high and the likelihood that the venture will continue to attract more sponsors in the South is high.

The opportunity to produce joint authored peer reviewed internationally refereed journals is equally growing with the opportunity of many types of collaborative ventures. We have been involved in the production of a book on ICT systems for Systems of Innovation building in China, India, Brazil, Thailand, South Africa and Southern Africa that brought authors from these countries and PhD candidates. This book has been recognized by the Open University which is using it for teaching, Harvard University, the World Bank and others. The interesting matter is that the authors come entirely from the South and the peer reviewers are the leading researchers on the economics of technical change in the North.

Another interesting dimension to the Globelics experiment is the attempt to develop regional networks such as the Africa network on the economics of innovation, learning and competence building, the Latin American network on the economics of innovation, learning and competence building and the Asian network of the economics of innovation, learning and competence building. Of these, ASIALICS has been running successfully for eight years now.

There have been attempts to create similar networks at the country level to network researchers in a given country and economy. For example, the formation and successful functioning of the Chinese Academy and Circles on the economics of innovation, learning and competence building (CICALICS); and the initiative to form a similar network in India. There have been also similar initiatives at University level. At the Tsinghua University (China), there have been successive doctoral academies that combined economics of innovation PhD Scholars with engineering PhD scholars sharing activities whilst following separate training. In addition, the Globelics Academy supported by Portugal and Finland has been running in Lisbon and Tampere alternatively where doctoral scholars have been invited at an equal ratio from Africa, Latin America, Asia and the rest of the world. This academy has generated not only networking but also improvement in the way PhD scholars upgrade the quality of their research and develop new and fresh theoretical and methodological insights and research linkages. In Africa we have initiated the African Globelics Doctoral Academy (AGDA)(http:\\agda.uonbi.ac.ke) and the plan is to run the doctoral training by moving it annually and by including some doctoral researchers and trainers from the world pool of established researchers.

6. A Model of Research Capacity Building?

At the First Globalics Conference, the book edited by Muchie, Gammeltoft and Lundvall, *Putting Africa First: The Making of African Innovation Systems* (2003), Aalborg University, Denmark, was launched in order to assist the output of Globelics. Subsequently, every year books have continued to be launched during the Globelics annual conference. Globelics and its associated regional and university academies certainly provide a model on how with minimal cost and with the engagement and self- motivated involvement and funding self-reliantly from local environment, research training with clear and tangible results and impacts cane be produced.

- The Governance of Globelics is simple. There is a scientific board that works with largely autonomous local organizing country researchers. This local force with peer review and advice from the scientific board produces the results that have been consequential every year.
- Research leadership has come from the scientific board, but this has not been imposed. It has been negotiated and seemed to elicit the right outcome in terms of research productivity.

- The dissemination of the research output follows the global network and the output is reviewed by the researchers from different parts of the world. Thus research utility also follows the global network route that has been established.
- The formation of different regional and university academies provides a focal point for a global engagement of researchers at the local sites of knowledge such as universities, country levels such as China and regional levels such as Africa, Latin America and Asia.
- Globelics is not a research NGO. It does not compete for funds or staff from the qualified pool of researchers. The research leadership is provided with open and inclusive style. This generates a voluntary communication that has continued to grow. There is debate in the board to formalise it or keep this open and voluntary approach to research leadership and communication.
- Globelics is mean to strengthen and complement the doctoral and post doctoral work of universities. It actually makes those from the universities and researchers to learn new research trends, to upgrade their capability, to engage in staff development, to acquire new networks, to learn of what the cutting edge research is, acquire new research and methodological and theoretical tools. It is thus an arm of universities rather than like the NGOs a contester to share, take or dilute the function and resources of universities.
- Globelics is self-organizing and inexpensive. This attribute is very important in relation to finding working models for enhancing and developing research capacity building initiatives that can stimulate research productivity and utility at the same time.

7. GLOBELICS and Universities

Universities are degrading in some places and upgrading in others. The Globelics initiative is one that can be built to assist universities in research capacity building. Globelics has opened the opportunity for Southern and Northern academics and researchers to interact at various levels. The first and foremost level is research capacity building. In the Globelics conferences, prior work involving researchers, and research interaction from many countries has resulted in the publication of peer reviewed articles, books and monographs. Often these publications tend to twin the best research leaders that initiated the Globelics venture with a variety of doctoral and post-doctoral scholars, researchers and academics. This has enhanced research training and research capacity building. In nearly all the Globelics conferences a number of books have been published and academics from the developing world (from Latin America, Africa and Asia) have had the opportunity to create new opportunities for networked learning, developing research projects on the developmental university.

Global initiatives for an international network of researchers combining every region of the world have been initiated. In such meetings opportunities existed to promote research networks such as resisting inequality through science and technology(RESIST), the Catch Up research project, the BRICS (research on systems of innovation in Brazil, India, China, South Africa and Russia) and others have been developed. More are likely to emerge through the researcher to researcher interactions.

One of the consequences (perhaps unintended) of the Globelics experiment is the degree to which knowledge power mal distribution may be contested by the sheer opportunity for many Southern researchers to open publishing where before they have little chance. Publishing and research is concentrated in the North, but access to Southern researchers through Northern researcher mediation facilitates being published in the best publishing houses and journals. This enhances and upgrades the capacity of Southern researchers.

The Globelics academies and the regional networks in Africa, Asia and Latin America complement the training and publication at a low cost budget. The real potential of these academies and networks has yet to be fully realized. Universities can enter into collaboration to enhance their own staff productivity. Their doctoral scholars can join the Globelics academy and acquire new methodological, theoretical and empirical strategies for the creation of knowledge. Other postgraduates can also be exposed to senior researchers from different parts of the world. This complements research capacity enhancement in relation to the work universities do. Though this might not mitigate the overall crises universities face or some now say the 'death of universities', it does assist universities to do a few things well.

What the Globelics initiative does is complement formal university education by creating the research environment and research leadership that can lead to substantial learning benefits to those engaged in the study of innovation in the context of development. Globelics (www.globelics.org) and country specific network such as Cicalics (www.cicalics.org) emerged as low cost, innovative and effective ways to upgrade scholarship in the field of innovation studies respectively in developing countries and the transition economies like Brazil, China, India and South Africa.

The Globelics model can be replicated and it can provide certainly, though incremental, a model of effective research capacity and cooperation where the result is directly measurable.

8. Concluding Remark

The policy relevance of the Globelics model is that it emerged from Northern researchers in collaboration and conversation with Southern researchers and has taken off largely as a self-organizing research capacity enhancing complement to university research activities. Bengt-Åke Lundvall, Richard Nelson, Luc Soete and Freeman from the North and Jose Cassialato, Shulin Gu and Mammo Muchie from the South originated the Globelics experiment with the overall leadership provided by Lundvall. It has produced an alternative way of how research capacity at low cost and effective results and output can be developed. It demonstrates a case study of how a research capacity initiative can be organized. Provided other fields in other areas uptake these supplementary type activities, there is a possibility the erosion of university research can be reduced if not eradicated.

This model is something that should not be ignored, but used to develop a mechanism for research capacity that is not heavily donorised. To be sure donors can incubate such low cost and efficient, result orientated research capacity initiatives. In a situation in many poor economies when the university is donorised, the state is donorised and even business is being supported by donors through private-led investment initiatives, it is critical that a research capacity device like Globelics with its academics, regional, country and university nodes- as an integrated research capacity cooperation model- emerges as a refreshing development. It shows provided researchers are willing to think clearly and commit themselves, the opportunity exists to take a number of initiatives to counter the crises of the universities without undermining the university as the premier idea, practice and institution of research and training. It is said nothing empowers as self-empowerment. The largely self-organizing and self-reliant approach of the Globelics experiment is worth replicating in stimulating complementary research capacity training by spreading informal networks of researchers to stem the decline of the university as quintessentially the research university. The re-emergence of the university as the knowledge and research centre can lead to the strengthening of a country's overall national system of innovation. Conversely, the strengthening of the system of innovation leads to strengthening of universities. This will stimulate a win-win outcome and is thus worth paying attention to alternative ways of delivering research capacity to re-centre the universities as the creators of knowledge capital for the future.

Notes

1 See: www.globelics.org

2 See particularly, World Bank (1988), *Education in Sub-Saharan Africa: Strategies for Adjustment, Revitalization, and Expansion.*

3 Leading work has been done by Chris Freeman (SPRU; UK), Richard Nelson (Columbia University, New York) and Bent Åke Lundvall (Aalborg University, Denmark) and others.

4 The 3rd GLOBELICS conference took place in Pretoria. Mammo Muchie was successfully involved along with Rasigan Maharajh and Simon Mpele from the DST in formulating the strategy for raising finance almost entirely from South African sources to organise the event. The same model appears to have been replicated in India and Malaysia subsequently. It remains to be seen whether other conference annually would replicate this practice or tradition of raising resources entirely from the local environment.

References

Archibugi, D. and Lundvall, B.Å. (2001), *The Globalising Learning Economy*, New York: Oxford University Press.

Baker, G. and May, T. (2002), 'Auditing as the Eternal Present: Organisational transformation in British higher education', in *European Political Science*, 1(3):12-22.

Castells, M. (1996), *The Rise of the Network Society,* New York: Blackwell.

Castells, M. and Hall, P. (1994), *Technopoles of the World*, London: Routledge.

European Commission (2000), *Communication from the Commission: Towards a European Research Area*, Luxumbourg: European Commission.

OECD (1996), *The Knowldge Based Economy*, Paris: OECD.

Soete, L. (2003), 'Speech at the First Globelics Conference', Rio: Brazil.

UNESCO (1998), *Conference Report on World Higher Education*, Paris: UNESCO.

World Bank (1971), *Education Sector Working Paper*, Washington DC: World Bank.

World Bank (1974a), *Education Sector Working Paper*, 11017, Washington DC: World Bank.

Word Bank (1980), *Education Sector Policy Review*, Pub 26-80, Washington DC: World Bank.

World Bank (1988), *Education in Sub-Saharan Africa: Strategies for Adjustment, Revitalization, and Expansion*, Washington DC: World Bank.

Ziman, J. (1994), *Promethus Bound: Science in a Dynamic and Steady State*, Cambridge: Cambridge University Press.

CHAPTER 6

Undergraduates' Access to E-learning in Nigeria:

Policy Implications

Nwaocha Vivian Ogochukwu

1. Introduction

The term e-learning is ambiguous to those outside the e-learning industry, and even within diverse disciplines it means different things to different people (Dublin, 2003). For instance, in companies it often refers to the strategies that the company network employs in delivering training courses to employees and lately in most Universities. E-learning is used to describe a specific mode of attending a course or program of study where the students rarely or never meet face-to-face, nor access on-campus educational facilities, because they study online. Simply put, e-learning is the acquisition of knowledge and skills using electronic technology such as computers and the internet-based courseware and local and wide area networks. It is a logical and strategic approach to achieve the technological transformation of Nigeria, thus the deployment of the

ICT is critical in the implementation of education roadmap, which is designed to revamp the education system. E-learning is an important platform for alleviating most of the problems currently facing the nation's education system.

The advent of information and communication technology (ICT) has led to an incredible revolution of Education in tertiary institutions all over the world. There has been a shift from the traditional approach of teacher-directed / didactic to modern methods where computer technology plays a significant role. Information and communication technology has promoted learning and made it more meaningful, where students can receive lectures from anywhere and at any time. The aspect of information and communication technology that has brought about this transformation in students' learning is e-learning.

According to Bugaje (2007), the 21st century is a knowledge century and prospective professionals to be produced in Nigeria must be current in knowledge. Unfortunately, the diminishing funding of tertiary institutions in Africa, coupled with increased students' enrolment has led to a decline in the quality of education (Ng'ambi, 2006). This problem seems most prevalent in Nigeria. Nwaka (2005) observed that as a consequence, Nigerian universities are now only a shadow of their former glory. The quality of scholarship and of the knowledge generated is constrained by the limited exposure of our scientists to current literature and modern techniques. This has resulted in their isolation from global discourse and trends in their fields.

In Nigeria, alertness to e-learning started gathering momentum two decades ago. The early exposure came through lecturers who studied abroad and had opportunities of attending conferences on e-learning technologies. Even at that, what was obtainable as was the case with University of Abuja was the lowest aspect of ICT such as print, audio/video tapes and digital radios (World Bank, 2002). This situation is similar in other universities in the country. This position is not a surprise because Nigeria has no specific policy for ICT in education. In February 2007, did the Federal Ministry of Education create its ICT department (WikiEducator, 2007). However, a veneer of e-learning exists, at the departmental levels, rather than institutional, and these departments are more in the medical, engineering, environmental sciences and computer science or informatics, where the synergy between research and teaching is strongest, and the essential infrastructure for course development and delivery are most accessible.

Similarly, synergistic opportunities have existed in research, marketing or programme development which stimulated involvement by schools of education, and by department of continuing education and extension studies (Curran and Fox, 1999). In these subject areas, students employ e-learning tools for their lectures. The slow pace in e-learning development may have in part accounted for the low rating of Nigerian universities globally in terms of impact and productivity in web-related activities. Interestingly, no Nigerian University was ranked among the first 50 in the world in terms of web size, papers, rich files and scholarship. In Africa, Nigerian University took the 44th position, while in the world it ranked 5,834 (InternetLab, 2007). According to Saint (1999), the core medium of instruction on the African continent remains print, with other technologies acting as a supplementary means of delivery. Nonetheless, there are significant investments which are being made in computer technologies to support functions such as the extension of programs to distant sites and the use of ICTs to support the teaching/learning, research and other management functions (LaRocque, 2003).

This scenario notwithstanding, government has in the recent past evolved and funded e-policy programmes aimed at providing e-learning facilities in tertiary institutions in Nigeria. This bold step was not sustained, and as such, further development of e-learning technology has attracted a not-so-good attention. This is mainly obtainable in universities. The situation at other levels of higher education is near blurred. Bugaje (2007) lamented that e-learning is very minimal in our curriculum; Internet connectivity in our institutions still relies on small cafes owned by private individuals, libraries depict the picture of museums with old and out-dated books, some of which existed since the colonial era, and there is no connection to virtual libraries or journal websites.

The University of the 21st century must embrace e-learning as a strategy for delivering quality teaching and offer services for its students (Shettima, 2005). Pushing further he maintained that e-learning includes digital library, computerization of administrative services, smart classrooms and e-journals. Universities could take advantage of new technological developments to address some of the problems of resource constraints, which is currently plaguing the university system in particular and higher education in general. For e-learning to succeed in Nigeria, it needs to build on another vital pillar, which is the existence of infrastructure, along with some degree of connectivity (Gunawardana, 2006).

NetTOM (2007) opined that cognitive gains from e-learning include hypertext learning which is non-linear and can be structured to engage learners into making greater use of critical thinking skills. Educational gains of e-learning include being forced to consider the requirements of learners and becoming more flexible with curriculum. Also, it enables learners to look towards teachers for perspective interpretation, analysis, motivation and guidance and teachers will expect learners to become more critical users of information and to generate their own contributions to knowledge.

It is on these issues that this paper aims at finding out the access to e-learning of students in higher education and the associated management implications.

2. Hypotheses

1. Students' access to e-learning equipment is not significantly low.
2. Students' exposure to e-lecture is not significantly low.
3. Students' access to training on e-learning technology is not significantly low.

3. Literature review

E-learning has been noted as the effective teaching and learning process created by combining e-digital content with local community and tutor support along with global community engagement. It is also an effective and efficient system of self-paced personal training, available over the internet (Open Learning Centre, 2007).

The Commission on Technology and Adult learning (2001) defined e-learning as instructional content or learning experiences delivered or enabled by electronic technology. Erah (2006) stated that the term e-learning refers to computer-enhanced training as opposed to the computer-based training of the 1980s. It is usually delivered in a personal computer and includes learning delivered by other communications technologies. According to him, e-learning is an approach to facilitate and enhance learning through both computer and communication technologies. The devices often used for this purpose include personal computers, CDROMs, television, personal digital assistants (PDAs – handheld devices that were originally designed as personal organizers, but became much more versatile over the years), MP3 players and mobile phones.

Communication technology enables the use of internet, e-mail, discussion forums, collaborative software, classroom management software, team learning systems, intranet, extranet, Local Area Network (LAN), Wide Area Network (WAN), audio and videotape, satellite and interactive television lectures, satellite-delivered learning, virtual educational networks, satellite downlinks, computerized diagnostic assessment, competency certification and electronic portfolios (Kaplain-Leiseron in Asogwa, 2006; The Commission on Technology and Adult Learning, 2001). It is facilitated and supported through the use of information and communication technology and is based on electronic formats (Bodys, 2005). E-learning can benefit any student irrespective of the background as a result of enhanced learning opportunities provided by ICT. A key issue in pedagogy is individualized that is, adapting the teaching to the needs of various learners (Asogwa, 2006). E-learning provides this by accelerating the required courses which eventually lead to increased graduation rates and lower drop-out rates of students (Olorundare, 2006). It promotes equity by providing students with access to courses which may not ordinarily be available (Asogwa, 2006).

NetTOM (2007) held that e-learning should be seen as offering solutions to several challenges currently facing higher education. These include the move towards lifelong learning, with its ongoing demand for continuous professional development, and the drive to wider participation. These challenges appear at a time of increasing pressure on resources and the increasing diversity in the student population and their modes of attendance, including learning that is part-time, at a distance open or flexible, and work based e-learning can improve the flexibility and quality of learning by:

(i) Creating an environment that promotes an active approach to learning.

(ii) Giving control to students over when and where they study.

(iii) Allowing students to study at their own pace.

(iv) Providing access to a range of resources and materials which may not otherwise be available or accessible for example graphics, sound, animation, multi-media.

(v) Providing a student with centered learning environment which can be tailored to meet the learning needs of individual students.

(vi) Creating an environment that promotes an active approach to learning.

(vii) Supporting increased communications between staff and students and amongst students.

(viii) Providing frequent and timely individual feedback, for example through computer assisted assessment and positive reinforcement.

(ix) Motivating students through appropriate use of interactive courseware.

(x) Supporting and encouraging collaborative learning.

(xi) Supporting economic reuse of high quality, expensive resources.

(xii) Encouraging students to take responsibility for their own learning.

There have been positive findings in distant learning programmes that use web technologies (Open Learning Centre, 2007). Several studies indicate that students taking distance learning courses have similar attitudes, grades and test scores as those students in traditional classrooms. There is a huge body of research on distance learning between the years of 1952 and 1992 that seems to conclude that learning outcomes are not at variance with those of a traditional classroom. The major problem with this research is that it looks primarily at televised instruction rather than web and computer based instruction. Another study found out that online students had more peer contact with others in the class, enjoyed it more, spent more time on class work, understood the material better, and performed, on average 20 percent better than students who were taught in the traditional classroom. Sadly, this advantage is yet to be of benefit to Nigerian students, as majority of them are computer illiterates. Folorunso, Ogunseye and Sharma (2006) found that mass unawareness, low computer literacy level and cost were identified as critical factors affecting the acceptability of e-learning by students and lecturers of Nigerian universities. In addition, access to connectivity to internet has remained one of the major challenges in many developing countries such as Nigeria. Darkwa and Mazibuko (2000) noted that students would need access to computers that can send and receive messages using web browsers.

Odusanya and Bamgbala's (2002) study revealed that the majority of final year students at the college of medicine, University of Lagos had limited computer skills and only 23 percent of the students had used the Internet for medical research. They concluded that the Nigerian student population is computer deficient. Similarly, Adomi and Anie (2006) reported that Nigerian industry asked universities to deepen the current exit knowledge levels of ICT skills of university graduates. According to them, industry was concerned about levels of graduate skills in online / Internet searching, systems management and computer keyboarding. However, despite this unsavory state of affairs in students' access to e-learning in universities, National Universities Commission (NUC), the government agency responsible for registering and regulating universities have prescribed personal computer ownership as follows: 1 Personal Computer (PC) to 4 students, 1 PC to 2 lecturers below the grade of Lecturer 1, 1 PC per Senior Lecturer and 1 notebook per Professor/Reader (Okhiria, 2007). Some universities like Nnamdi Azikiwe University have met this mark in this PC-lecturer ratio, but are unable to meet that of students. Some universities have equally made giant strides in campus wide area networking and e-learning course deliveries (WikiEducator, 2007).

In contrast of the Nigerian situation to that of South Africa, a study of the qualifications of employees in that country's ICT industry revealed that there are very few post graduates in the technical group of employees. This poses the question of whether the ICT industry had developed to a level where the services of post graduates are not required or whether the current employees are just under qualified. (Vukanikids – DTI, 2005). LaRocque (2003) reported that Botswana, Cameroon and Zambia are using a university-based internet system to support interactive regional study centres for distance learners, while Tanzania, Botswana and Zimbabwe have established new tertiary institutions wholly dedicated to distance education. Furthermore, in a study (Steiner *et al.*, 2003) using 100 universities in 50 countries in Africa, it was found that no Nigerian university ranked among the top ten or the bottom ten nor the middle position in bits per networked computer. Similarly, they provided information on total bandwidth requirements in the sampled Nigerian universities (see Table 1 and Figure 1).

Table 1: Approximate Value of Total Bandwidth Requirements in Surveyed Tertiary Institutions in Nigeria

Institutions	User Size	Total No of Networked Computers	Total Mbps Available	Mbps Required	Band-width Difference
Obafemi Awolowo University, Ile-Ife	30950	1000	0.77	22.22	21.45
University of Ibadan, Ibadan	27300	805	1.26	17.89	16.63
Ahmadu Bello University, Zaria	29400	450	0.64	10.00	9.36
University of Port Harcourt, Port Harcourt	15000	300	0.38	6.67	6.29
University of Jos, Jos	26150	300	0.77	6.67	5.90
Bayero University, Kano	32224	178	0.19	3.96	3.77
Rivers Satate University of Science and Technology, Port Harcourt	25400	100	0.13	2.22	2.09
University of Abuja, Abuja	11010	-	0.19	-	-

Figure 1: Chart Depicting the Estimated Total Bandwidth Requirements in Surveyed Tertiary Institutions in Nigeria

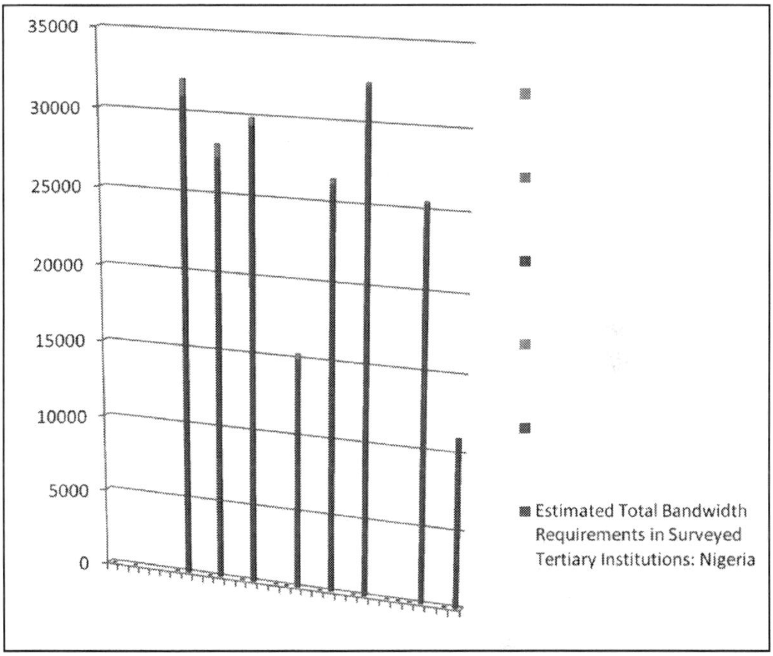

A strategic study carried out on behalf of the EU commission showed that while the level of integration of ICT in teaching has increased greatly over the last two years, considerable variation still exists between institutions in this regard (Ramboll, 2004). However, despite the variations, e-learning activity in Europe continues to grow. A survey of the use e-learning in training and professional development showed that some 14 percent of total spending by users of training went to e-learning related content in 2001 – "appreciable more than two years earlier when the figure was under 10 per cent" (CEDEFOP, 2001:48). Furthermore, there is abundant evidence to show that widening access to e-learning programmes and to their related academic resources is an important objective of many university e-learning strategies, reaching new groups of students is an additional and closely related goal (Waits *et al.*, 2003).

A commitment to widening access to education is a common

goal of many university e-learning strategies. In a survey on the importance of various goals to institutions; distance education programmes (a high proportion of which use online technology as a primary or supporting medium of instruction) 2 out of 3 U.S. four-year public institutions indicated that increasing students' access was a very important goals; either by making courses available at convenient locations (72 per cent) or by reducing time constraints for course taking (66 per cent). The corresponding figures for four-year private institutions were also high (65 and 61 per cent respectively). Similarly, 69 per cent of four year public institutions indicated that increasing the institutions' access to new student audiences was an important goal of their e-learning programmes; the corresponding figure for four-year private institutions was 64 per cent. Thus, increasing institution enrolments was ranked quite high: 58 per cent for four-year public institutions and 57 per cent for the same year private institutions (Waits *et al.*, 2003). Studies in e-learning have shown that most programmes are likely to succeed with the constant involvement of the facilitator through e-mail discussion lists and individualized massages (Gunawardana, 2005). According to him, the facilitators constant involvement and feedback are the most important support for the learner.

Open Learning Centre (2007) found that e-learning produces higher retention of content through personalized learning. According to them, students' can customarize the learning material, to their own needs and as such have more control over their learning process and can better understand the material, leading to a 60 per cent faster learning curve, compared to instructor-led training. The delivery of content in smaller units called "chunks" contributes further to a more lasting learning effect. Whereas the average content retention rate for an instructor-led is only 58 per cent, the more intensive e-learning experience enhances the retention rate by 25-60 per cent.

4. Methodology

The main thrust of this work is on higher institutions in the Geo-political zones of Nigeria consisting of Universities, Polytechnics and Colleges of Education. The design adopted for this study was ex post facto. Three hypotheses were formulated to give direction for the execution of this study. Undergraduates in tertiary institutions formed the population of this study. Based on the aims of the study, the sample size consisted of 500 students, drawn from five higher institutions using stratified random sampling technique. The institutions were 2 universities (1 federal and 1 state owned), 2 polytechnics and 1 college of education. 100 students were randomly drawn from each university while 100 students were drawn from each polytechnic and college of education.

Data for analyses were collected with researchers constructed instrument called Students' Access to E-learning Technology Inventory (S.A.E.T.I.). It had two sections – A and B. Section A contained demographic variables which were seven in number. Section B contained 35 items on a 10-point rating scale. 12 each items measured 2 variables while 11 items measured the remaining 1 variable. The instrument was validated by experts in measurement and evaluation, while the trial test yielded reliability coefficients ranging from 0.70 to 0.89. These figures were deemed good enough to measure consistently what the instruments purports to measure over a given period of time.

The administration of the instrument was personally handled by the researchers and in some areas with the aid of research assistants. 500 copies of the instrument were administered to the sampled students and collected back. This gave a 100 percent return rate.

Data collected were subjected to statistical analysis using descriptive statistics and population t-test analysis for single means.

5. Results

1. Students' access to e-learning is not significantly low. Using the population t-test analysis for single means, this hypothesis was tested. The result of the analysis is presented in Table 2.

Table 2: Population t-test Analysis for single means of level of Undergraduates' Access to e-learning Equipment (n = 500)

Variable	**Expected**	**Observed**	**Standard**	
	Mean u	$\bar{X}$	*SD*	*T*
Laptops	5.50	2.32	2.06	-34.54*
Overhead projectors	5.50	2.43	2.12	-34.42*
Slide projectors	5.50	2.45	2.43	-28.06*
Powerpoint projectors	5.50	2.59	2.41	-26.95*
Multimedia projectors	5.50	2.42	2.21	-31.22*
Data projectors	5.50	2.30	2.16	-33.11*
One-way interactive equipment	5.50	3.11	2.83	-18.92*
Two-way interactive Equipment	5.50	3.22	2.91	-17.57*
Local Area Network (LAN)	5.50	3.09	2.91	-18.59*
Wide Area Network (WAN)	5.50	3.61	3.14	-13.49*
Classroom based computers	5.50	3.02	2.79	-19.85*
Home computers	5.50	3.01	2.69	-20.71*
File server computers	5.50	4.33	3.41	-7.67*
Library based computers for student instruction	5.50	4.48	3.24	-7.66*

* Significant at .05; df = 499; critical t = 1.96

Table 3: Population t-test Analysis for single means of level of Students' Exposure to E-lecture (n = 500)

Variable	Expected Mean	Observed Mean	Standard Deviation	
	U	$\bar{X}$	*SD*	*t*
Access to telecommunication services	5.50	4.70	3.24	-5.51*
Access to telephone (Teledensity)	5.50	3.12	2.65	-20.05*
Access to internet connection (networking)	5.50	2.74	2.58	-23.90*
Access to hypertext learning	5.50	2.54	2.52	-26.20*
Use of internet at school library	5.50	3.17	2.79	-18.66*
Use of internet as part of classroom instruction	5.50	4.40	3.40	-7.21*
Use of internet as part of school course work	5.50	3.33	2.95	-16.42*
Use of internet by students for research	5.50	3.11	2.63	-20.30*
Use of electronic text books, copyright and e-learning method related topics	5.50	2.86	2.54	-23.27*
Lecturer guidance on the use of e-learning technology	5.50	2.82	2.37	-25.32*
Instructors guidance on the use of e-learning technology	5.50	2.87	2.45	-24.05*

* Significant at .05; df = 499; critical t = 1.96

The result presented in Table 2 showed that students access to e-learning equipment is significantly low in respect of laptop (t= -34.54), overhead projectors (t= -32.42), slide projectors (t= -28.06), power-point projectors (t= -26.94), multi- media projectors (t= -31.22), data projectors (t= -33.19), one-way interactive equipment (t= -18.92), two-way interactive equipment (t= -17.57), Local Area Network (t= -19.59), Wide Area

Network (t= -13.49), Classroom based computers (t= -19.85), home computers (t= -20.71), File server computers (t= -7.67) and library based computers for students' instructions (t= -7.06). The null hypothesis that predicted that student access to e-learning equipment is significantly high was rejected. Further observation of Table 2 showed that the observed mean access to e-learning equipment by students was lower for all equipments than the expected mean access to e-learning equipment of 5.5. When these observed mean values were compared to the expected mean value of 5.5 using the population t-test analysis for single mean, negative t-test values were obtained. This finding implies that students' access to e-learning equipment is significantly low. In other words, not many students have had access to the e-learning equipments enumerated in the study.

2. Students exposure to e-lecture is not significantly low. The population t-test analysis for single mean was done to test this hypothesis. The result of the analysis is presented in Table 3.

   The result presented in Table 3 revealed that the students exposure to e-lecture is significantly low given access to telecommunication services (t= -5.51), access to telephone penetration (t= -20.05), access to internet connections (t= -23.90), access to hypertext learning (t= -26.20), use of internet at school library (t= -18.66), use of internet as part of classroom instruction (t= -7.21), use of internet as part of school course work (t= -16.47), use of internet by students for research (t= -20.30), use of electronic textbooks, copyright and e-lecture related topics (t= -23.27), lecturers guidance on the use of e-learning technology (t= -25.32), instructors guidance on the use of e-learning technology (t= -24.05). The null hypothesis was rejected because the calculated t-values were found to be higher than the critical t-value of 1.96 given 0.05 alpha value and with 498 degrees of freedom. Further examination of Table 3 showed that the expected mean value of 5.5 was found to be greater than the observed mean values in respect of the variables. Statistical comparison of the expected and observed mean values for each variable using the population t-test analysis gave significant t-values.

The finding means that students' access to e-lecture was significantly low. In other words, students are not exposed to e-lecture

3. Students' access to training in e-learning technology is not significantly low. Using the population t-test analysis for single mean this hypothesis was tested. The result of the analysis is presented in Table 4.

The result presented in Table 4 showed that students access to training in e-learning technology is significantly low with respect to easy access in operating e-learning equipment (t= -25.97), technical difficulties in operating e-learning equipment (t= -32.88), handling multi-media software (t= -41.63), departmental sponsorship to training on e-learning technology (t= -36.49), faculty sponsorship to training on e-learning technology (t= -31.68), school sponsorship to training on e-learning technology (t= -23.64), non-governmental organizations sponsorship to training on e-learning technology (t= -28.94), school provision for purchasing or renting new computers (t= -28.19), faculty provision for purchasing or renting new computers (t= -18.92) and departmental provision for purchasing or renting new computers (t= -19.85).

The negative t-values means that the observed mean values for the variables were lesser than the expected mean value of 5.5. The null hypothesis was thus rejected because the calculated t-values were higher than the critical t-values of 1.96 given .05 alpha level and with 498 degrees of freedom. This finding means that students' access to training in e-learning technology is significantly low. In other words, students do not really have access to training in e-learning technology.

Table 4: Population t-test Analysis for single means of Students' access to training on e-learning technology (n = 500)

Variable	Expected Mean	Observed Mean	Standard Deviation	
	U	*X*	*SD*	*T*
Easy access in operating e-learning equipment	5.50	2.68	2.43	-25.97*
Technical difficulties in operating e-learning equipment	5.50	2.23	2.22	-32.88*
Training in handling multimedia software	5.50	2.02	1.87	-41.63*
Departmental sponsorship to training on e-learning technology	5.50	2.18	2.03	-36.49*
Faculty sponsorship to training on e-learning technology	5.50	2.30	2.56	-31.68*
School sponsorship to training on e-learning technology	5.50	2.81	2.54	-23.64*
Non-governmental organizations sponsorship to training on e-learning technology	5.50	2.55	2.28	-28.94*
School provision for purchasing or renting new computers	5.50	2.55	2.34	-28.19*
Faculty provision for purchasing or renting new computers	5.50	3.10	2.82	-18.92*
Departmental provision for purchasing or renting new computers	5.50	3.01	2.75	-19.83*

* Significant at .05; dt = 499; critical t = 1.96

6. Discussion

The outcome of hypothesis 1 revealed that in Nigeria, undergraduates' access to e-learning is significantly low. In other words, the availability of e-learning equipment such as laptops, overhead, slide, power-point, multi-media and data projectors, one and two-way interactive equipment, Local Area Network (LAN), Wide Area Network (WAN); classroom, home, file server and library-based computers for students and access to them were found to be low.

The outcome of this study confirms Odusanya and Bamglala,

(2002), Idowu *et al.* (2004), and Adomi and Anie (2006) findings. Consequently, Nigerian universities have not been able to reap the benefits of the provision of expanded access without incurring additional infrastructure costs, reducing the opportunity costs of students' study, reduction of institutional budgets dedicated to staff salaries, distance education which enhances the decline in cost per student over some range of output, and the updating and modification of e-learning materials in an incremental manner (LaRocque, 2003).

The results of hypothesis 2 indicated that Nigerian students' exposure to e-lecture is significantly low. By implication, students' access to telecommunication services, telephone penetration, internet connections, hypertext learning, internet use at library, classroom, coursework, students' research, use of electronic textbooks among others is at a dismal level. This study's outcome is in line with the works of Folorunso *et al.* (2006), and Ng'ambi (2006). However, it is at variance with what is obtainable in Botswana, Cameroon, Zambia, Tanzania and Zimbabwe where university-based Internet system is used to support interactive regional study centres for distance learners (LaRocque, 2003). The possible explanation for this finding centres around low percentage of lecturers with ICT skills, lack of requisite telecommunications infrastructure capable of transporting multi-media messaging, uneasy access to computer equipment and other accessories at institutional and personal levels due to the location of cyber-cafes in commercially profitable urban areas, to the detriment of rural and semi-urban areas. All these among others constitute constraints to ICT usage and the facilitating of e-learning in education (WikiEducator, 2007).

However, there are some pleasant developments in other areas within the country. The Obafemi Awolowo University for instance, located in the South West has made impressive gains in ICT. It has the best developed ICT system in Nigeria with its own VSAT access to the Internet and campus-wide Intranet. Similarly, National Open University of Nigeria has also achieved some remarkable feats in e-learning while Nnamdi Azikiwe University in South-East has been able to provide personal computers to lecturers. Furthermore, the University of Jos located in North-Central is blazing the trail for content development and e-learning in addition to campus networking (WikiEducator, 2007).

Another significant outcome of this study (hypothesis 3), was that university students in Nigeria's access to e-learning technology is significantly low with respect to easy access to operating e-learning equipment, technical difficulties in operating e-learning equipment, training in handling multi-media software; departmental, faculty, school and non-governmental provisions for purchasing or renting new computers. This study's outcome fell in line with the report on the conference on African Research and Education Network Infrastructure (AAU, 2005). Offering explanation for this finding, the inability of technical manpower to transform their wealth of knowledge and experience into training students to acquire the basic skills on ICT has stunted students' access to training on e-learning technology. Another cogent reason for this finding hinges around the failure in establishing a Nigerian Research and Education Network (NREN) due to the fact that the focus was on NREN creation rather than evolution, the process was not organic with project schedules not significantly thought through, and too much emphasis on money, equipment and building, rather than on human capacity and social engineering.

However, to redress this ugly trend, Okojie (2007) disclosed that National Universities Commission has provided equipment and training on e-learning project for the University of Lagos as a pilot centre. According to him, NUC has also trained several lecturers online on how to improve their didactic skills, modern methods of teaching and learning, among others.

7. Conclusions

The University administration should go into public and private partnership such as the one existing with AfriHUB, Socket Works and should expand to include NIIT. By bringing in private sectors, the universities can meet their e-learning objectives, through the provision of cost-effective means of increasing access, equity, and quality in both formal and vocational education.

Universities should put in place an effective ICT board that will liaise with them to oversee the implementation and facilitate increase access to capital, introduce market discipline in decision-making, introduction of private sector knowledge and management skills to public sector. The university administration should also support private sector e-learning initiatives. This should include the provision of e-learning instruction for pre-service and in-service professional development of teachers and training of students and contracting with the private sector in financing, equipping and maintaining the computer laboratories in the universities.

The University administration through their liaison offices should liaise with multi-national agencies such as the USAID, who provide the Digital Opportunity through Technology and Communication for a five-year period which involves $15 million U.S. government effort to promote information and communication technology for development.

Universities should form a consortium to purchase bandwidth. This will enable them purchase a greater volume which will lower the marginal cost of that bandwidth. The implication of this is that where small groups of universities came together, they can dramatically lower their per –unit bandwidth costs.

The University administration in conjunction with the government should put in place modalities to enable undergraduate students to acquire their personal computers. This will not only alleviate the problem of computer literacy, but also enhance students' access to them.

Hence, in order to reposition the Nigerian educational system to meet global standards, and to fast-track economic development, access to e-learning at all levels of tertiary education should be accorded priority.

References

Adomi, E. E. and Anie S.O. (2006), 'An assessment of computer literacy skills of professionals in Nigerian university libraries', in *Library HiTech News,* 23 (2): 10-14.

Association of African Universities (AAU) (2005), *Report of the conference on African Research and Education Network Infrastructure.* November 14-15, Accra: AAU.

Asogwa, U. D. (2006), 'E-learning: A panacea for access, equity and quality higher education in Nigeria', *Paper presented at the 30th annual conference of Nigerian Association for Educational Administration and Planning (NAEAP),* The Faculty of Education Hall, Enugu State University of Science and Technology, October 2-3, Enugu.

Bodys, W. (2005), *The integration of information communications technology in Scottish schools*, An interim report of Her Majesty's Inspectors of Education. Accessed: http://www.hmie.gov.uk/documents/publication/EvICT%20Final%2018%20Oct.html

Bugaje, I. (2007), *Nigeria: Education Reform and Muslims*, March 19. Available at: http://www.allafrica.com (Accessed: 18 April 2007).

CEDEFOP (2001), *E-learning and training in Europe*, CEDEFOP Ref. Series 26, Luxembourg: Office for official publications of the Europe Communities.

Curran, C. and Fox, S. (1999), *Telematics in open and distance learning*, Weinheim: Deutscher Studied Verlag.

Darkwa, O. and Mazibuko, F. (2000), *Creating Virtual learning communities in Africa: Challenges and prospects*. Available at: http://www.firstmonday.dk/issues 5.5/darkwa (Accessed: 11 May 2007).

Dublin, L. (2003), 'If You Only Look Under the Street Lamps... Or Nine e-Learning Myths', in *The eLearning Developers' Journal*, June 10, pp. 1-4.

Erah, P.O. (2006), 'Introduction to e-learning protocols', *Paper presented at ETF capacity building workshop for lecturers of Universities in Nigeria*, The University of Uyo, September 26, Akwa Ibom.

Folorunso, O., Ogunseye, O. S. and Sharma, S. K. (2006), 'An exploratory study of the critical factors affecting the acceptability of e-learning in Nigerian universities' in *Information Management and Computer Security Journals,* 14 (5): 496-505.

Gunawardana, K. D. (2006), 'An empirical study of potential challenges and benefits of implementing e-learning in Sri Lanka', in *International Journal of the Computer, the Internet and Management*, Special Issue-August.

Idowu, B., Adagunodo, R. & Idowu, B. (2004), 'Gender differences in Computer literacy among Nigerian undergraduate students', in *The African Symposium: An online Educational Research Journals*, 4 (3): 1-7.

InternetLab (2007), *Webometrics ranking of world universities*. Available at: http://www.webometrics.info/comparative-productivity.asp (Accessed: 24 April 2007).

Kaplain-Leiseron in Asogwa, U. D. (2006), 'E-learning: A panacea for access, equity and quality higher education in Nigeria', *Paper presented at the 30th annual conferences of Nigerian Association for Educational Administration and Planning (NAEAP)*, The Faculty of Education Hall Enugu State University of Science and Technology, October 3-6, Enugu.

LaRocque, N. (2003), *The promise of e-learning in Africa: The potential for public-private partnerships*. Available at: http://www.businessofgovernment.org/pdfs/LaRocqueReport.pdf (Accessed: 24 April 2007).

NASBE in Asogwa, U. O. (2006), 'E-learning: A panacea for access, equity and quality higher education in Nigeria', *Paper presented at the 30th Annual Conference of Nigerian Association for Educational Administration and Planning (NAEAP)*, The Faculty of Education Hall Enugu State University of Science and Technology, October 3-6, Enugu.

NetTom (2007), *Why is e-learning important*? Available at: http://ebdd.wsu.edu/eder/NetTom_ToT/unit.1/whyelearning.htm (Accessed: 25 February 2007).

Ng'ambi, D. (2006), *ICT and economic development in Africa: The role of higher education institutions*. Cape Town: Centre for Educational Technology, University of Cape Town.

Nwaka, G. I. (2005), 'Higher education, the social sciences and material development in Nigeria', *Paper presented at 11th General Assembly of the Council for the Development of Social Science Research in Africa (CODESRIA)*, Maputa, Mozambique. Available at: http:www.codesria.org/Links/conferences/general_assembly11/papers /nwaka.pdf (Accessed: 13 May 2007).

Odusanya, O. O. and Bamgbala, O. A. (2002), 'Computing and information Technology skills of final year medical and dental students at the College of Medicine, University of Lagos, Nigeria', in *Nigerian Post Graduate Medical Journal,* 9 (4), 190-193.

Okhiria, P. (2007), *ICT tertiary level education: Universities and colleges of education.* Available at: repairs. dochttp://www.wikieducator.org/ICT4Africa/country ReportNigeria (Accessed: 19 February 2007).

Okojie, J. A. (2007), *ICTs for education development and empowerment: Children and learning for learning.* Available at: http://www.digitalworldafrica.org/icts. 4education_okojie.doc on (Accessed: 11 May 2007).

Olorundare, S. (2006), 'Utilization of ICT in curriculum development, implementation and evaluation', *Paper presented at the National Conference on Information and Communication Technology in service of education, Institute of Education*, University of Nigeria, May 15-18, Nsukka.

Open Learning Centre (2007), *What is e-learning*? Available at: http://www.openlearningcentre.com/what is e-learning.htm (Accessed: 25 February 2007).

Ramboll, P.L.S. (2004), *Studies in the context of the e-learning initiative: Virtual models of European Universities*, Draft final Report to the E. U. Commission, Copenhagen: DG Education and Culture.

Saint, W. (1999), *Tertiary distance education and technology in sub-Saharan Africa.* Washington D.C.: The World Bank.

Sedov, D. (2004), 'Wither education in Nigeria?' in *Worker's Alternative,* July-August, 1-3. Available at: http://www.google.com (Accessed: 12 May 2007).

Shettima, A. K. (2005), *Nigerian university education in the 21st century.* Available at: http://www.macfound.org/site/c/ (Accessed: 20 February 2007).

Steiner, R., Tirivayi, N., Jensen, M. and Gakio, K. (2003), *Enhancing Africa Research and education networks*. Available at: http://event-africa-networking.web.cern.ch (Accessed: 10 May 2007).

The Commission on Technology and Adult Learning (2001), *A vision of e-learning for Americas workforce*, Report of the Commission on Technology and Adult Learning. Washington, D. C.: ADSL/NGA.

Vukanikids-DTI (2005), *An analysis of the ICT skills audit for the DTI and ISETT SETA*, Department of Trade and Industry (DTI), South Africa. Available at: http://www.isett.org.za/download/ICT-skills-needs.pdf (Accessed: 11 May 2007).

Waits, T., Lewis, L. and Greene, B. (2003), *Distance education at degree-granting post secondary institutions: 2000-2001. NCES 2003-017*, Washington, D. C: National Centre for Education Statistics.

WikiEducator (2007), *ICT in education in Nigeria*, ICT4Africa/country Report Nigeria. Available at: http://www.wikieducator.org/ ICT4Africa/country Report Nigeria (Accessed: 11 May 2007).

World Bank (2002), *Enhancing learning opportunities in Africa*, Washington, D.C: International Bank for Reconstruction and Development.

CHAPTER 7

Demystifying and Demythologizing Technology and Technology Education

Thabo I Pudi, Vuyisile Msila, Cheryl le Roux, Agnieta B Pretorius, Coneth Richards & Marius Dieperink

1. Introduction

Michael Dyrenfurth (Daugherty and Wicklein, 2010:2), commented: "While TE is considered an essential characteristic of quality education, there are often misinterpretations and misrepresentations associated with it (TE)." This is exacerbated by the inconclusive agreement in either policy or practice over the definition of *Technology* and TE (DoE, 2009:14). It is partly because of this, that Technology and emerging technologies are *perceived* as mysteries (O'Donovan, 2009:1). In agreement with Brown and Brown (2009:8), TE as defined by ITEA may not be consistent with the practicalities of the classroom. Because of its newness in the curriculum, many teachers perceive it as a mystery (Daugherty and Wicklein, 2010:2). This is demonstrated by its implication in the two curriculum reviews that were undertaken by the department of education in 2000 and in 2009. In the first review headed by Linda Chisholm (DoE, 2000), a drastic recommendation was made to delete TE from the curriculum. This recommendation was reversed by the then education minister Prof. Kader Asmal (DoE, 2000). This reversal was hinged on the belief that TE is of vital need, especially for a growing country like South Africa. This notion is corroborated by Moodie (2009:1) that "TE has outcomes that South Africa should cherish rather than disparage".

In 2009, a commission was established to investigate the implementation of the curriculum through the National Curriculum Framework (DoE, 2009). Once again TE was implicated as one of the obstacles. Consequently, a recommendation was made that TE in the intermediate phase be subsumed into Natural Sciences. This prompted an outcry particularly from some Non Governmental Organizations (NGO's) and other TE interest groups that the recommendation to merge TE into Natural Sciences in the Intermediate Phase is a clear depiction of lack of understanding of what TE stands for and what it really entails. According to Moodie (2009:1) the Report has misread the purpose and outcomes of TE.

Dizzying advances in technology (O'Donovan, 2009; and Dryden and Vos, 2001) are in part responsible for the difficulty of the implementation of the TE curriculum. The nature of technology and TE together with technical terms used adds to its implementation peril. Most concepts that are used in the practice of TE are new. Sometimes known concepts are used but with a different interpretation and understanding. It is these interpretations of the concepts used that make TE a problematic subject.

Although TE in the South African context is a stand alone learning area, some of the concepts used in the TE curriculum straddle the entire National Curriculum where the emphasis with the Natural Sciences is more pronounced. For example, Brown and Brown (2009:1) points out the predominant confusion of TE with Educational Technology. Brown and Brown (2009:49) believe that the demystification of TE can be done in part by delving into what the contents of the 'official curriculum' stipulate. These include "the actions that take place inside the classroom, teaching and learning;

specific curricular material; transactions and process categories, as well as assessment."

Despite the fact that Technology is as old as mankind it has little or no established history as a learning area (Pudi, 2002:44; McCormick, 1997: 137). This is corroborated by Warner, Elrli, Johnson and Greiner (2007: 71), whose studies emphasized the lack of research efforts being conducted in TE. Because of this lack of history for TE, there is limited knowledge of what TE is about, that is, its aims, rationale, content and curricula (Daugherty and Wicklein 2010:1; Ankiewicz, 1995:2). It is this limited knowledge that has given birth to misunderstandings, misinterpretations and misrepresentations of what Technology and TE really are. It is easy to mythologize Technology and TE if we do not know their parameters. Knowing and understanding TE and its role in education is one way to demystify and demythologise it. TE will be demystified if it is given its rightful place within the curriculum. With some efforts, the demystification will lead to the demythologization. But first we have to understand clearly what is meant by mystery and myth and how these can influence the understanding of technology and TE. This is a logical step towards demystifying and demythologizing Technology and TE.

2. Myths about "Technology" and "TE"

Often myths emanate from true or real situations. This is exacerbated by the fact that truth itself is relative (Hamm, 1993:65; Jeevenantham, 1999:118). What may be true today could be a myth tomorrow. Hamm (1993:65) makes an example that once the earth was thought of as being flat and that was the truth. Today, theories that purport the earth to be flat are myths.

There are many myths surrounding TE. For example, there is reference to the myth that TE is a subject which is only good for learners who cannot cope with pure academic work (Pudi, 2009:1). As such, it is also related to handwork or the crafts (Hodgkin, 1990:208). Another myth is that as a result of its perceived emphasis on handwork or manual labour, (TE) is thought not to be suitable for women. Thus TE has been associated with gender roles.

According to Moghaddam (1998:427), gender or the social construction of society shows much prejudice towards women. The seriousness of this is seen in describing careers such as technological and scientific careers as suited for males only (Gaganakis, 1999:148 & Anon, 1996:15).

One of the mysteries surrounding the subject of TE is that it is often thought of as a subject requiring mathematics or science as a prerequisite. This myth has been advanced as the reason that many girls are unwilling to take TE at school. According to Moodie (2009:1):

> Technology as a school subject does not have the familiarity that Maths and Science enjoy in the minds of the public. While people feel that they know what school maths and science involve, Technology in many people's minds is associated with "learning about computers" or hand-crafts. In school it is often regarded as a time-table filler which requires no specialised knowledge or skill.

The myths about Technology and TE stem from ignorance, a lack of understanding, hasty conclusions and a distortion of the truth (Rudinow and Barry 1984:165). Myths also emanate from hearsay, cover-up, stereotype, lack of knowledge and understanding, jokes and sometimes serendipitously.

The presence of myths has the potential to inhibit learning and teaching about Technology and TE. Lewis (1999: 5), believes that the understanding of misconceptions and myths about Technology and TE is a prerequisite for better teaching and learning.

Myths are easy to acquire but difficult to get rid of. Like old habits, myths are hard to die. They spread like wild fire because they are logical and interesting (sensational). To acquire a myth is easy. The problem lies in trying to unlearn the myth. According to Steinem (2008:1), unlearning is a lot more difficult than learning. The art of unlearning something is difficult. Besides being logical, myths are also fun. Myths can emanate from exaggerations, misinterpretations such as misinterpreting the media adverts, making jokes or sarcasm or simply telling a lie.

3. Technology and Technology Education (TE)

Daugherty and Wicklein (2010:2), indicates that there is general confusion over what Technology and TE really entail. Consequently, there is a misunderstanding of the concepts "Technology" and

"Technology Education". The question of what Technology and TE really are has been debated since the 1980s (Pullias, 1992:3). Kramer (1994:2), maintains that Technology is the blending together of everything that one knows and can do to solve specific problems. Maley and Wenig (Daugherty and Wicklein, 2010:2), point out that there is still considerable confusion over what characteristics exemplify TE today. This undoubtedly has a negative impact in the implementation of TE. For effective implementation, any confusion or misunderstanding and half-truths about Technology and TE should be cleared up in an investigation involving definitions and explanations of these concepts.

The concept "technology"

The definition of Technology is broad. Technology can be defined in terms of what we need to achieve with that definition. Different authors define technology in different ways. Kramer (1994:1) concedes that there is no single correct definition of technology. This is because most of them are to some extent true as definitions of Technology. According to Hodgkin (1990:208), technology is the words or theory about the *techne* which is art and craft or the making of "good things". 'The term "Technology" is derived from the Greek words **tekhno**, which refers to an art or craft, and **logia**, meaning an area of study. Thus, according to Hodgkin (1990:208), technology means the study or science of crafting'.

The Technology Learning Area Workshop Committee Report (DoE, 1996:1) defines technology as 'a process using knowledge, skills and resources to meet human needs and wants; to recognize and solve problems by designing, developing and evaluating products, processes and systems'. The Revised National Curriculum Statement (RNCS) (DoE, 2002:14) adds values, skills and knowledge to this definition and emphasizes sensitivity to social and environmental factors. Cross, Naughton and Walker (1986:27), refer to technology as "the application of scientific and other organized knowledge to practical tasks by social systems involving people and machines".

The definition of technology from Volti (1988:3-6); DoE (2002:4); Black (1998:1); Williams (1992:1); Dugger and Yung

(1995:1); Shield (1996:3); DoE (1997:89) and Ort-Step Institute (1995:5) conforms to the definition as given by The Technology Learning Area Committee Workshop (DoE, 1996:1) above and will not be repeated here.

The concept "Technology Education"

TE can be seen as a comprehensive experience-based educational programme that allows learners to investigate and experience the means by which people meet their needs and wants, solve problems and extend their capabilities. According to Texas Education Agency (2010:1), TE is "concerned with the knowledge and skills necessary to develop, produce and use products or services, and how to assess the impact of these activities on humanity and the environment." While it is clear that TE prioritizes the effective use of technological products and systems, it should not promote gender bias (Grimsley, 1999:1; Van Rensburg, Myburgh and Ankiewicz, 1996:2; Treagust and Mather, 1990:53; Hansen, 1999:244-296).

Grimsley (1999:1) defines TE as "a comprehensive experienced-based educational programme that allows students to investigate and experience the means by which humans meet their needs and wants, solve problems, and extend their capabilities." According to Potgieter (1998:3), TE is concerned with the technological knowledge and skills, as well as technological processes. This involves understanding the use of technology and its impact on both the individual and society. It is ultimately designed to enable and equip learners to perform effectively in the technological environment in which they live and to stimulate them to contribute to its improvement.

4. Technology Education in context

The aim of the curriculum is according to the constitution that of nation building. A country in transformation has a dire need for skills especially the skills to create a viable economy. Technology in general is an engine house for the development and prosperity of any nation in terms of the historical, cultural, social and the economic perspectives (Potgieter, 1998:13; Pudi, 1999:1; Poon, 1994:16). As such, efforts to establish TE as a learning area, especially for a developing country like South Africa cannot be overemphasized.

However, due to the perceived congestion of the curriculum and the change in the education paradigm, TE was wrongly thought of as an excess learning area. This was confirmed by the recommendations of the review committee (Chisholm, 2000). Amongst others, the review report depicted the congestion of the curriculum; the complaint of overload by teachers and the negative attitudes of teachers and learners towards TE. The negative attitudes according to the report emanated mostly from the lack of understanding of what Technology and consequently TE entails. This lack of understanding will bring about misconception and misrepresentation of technology and TE (Daugherty and Wicklein, 2010:2). The newness of TE within the South African curriculum and its simultaneous inception with the new education paradigm – the Outcomes-Based Education paradigm are possible springboards for the mystification and mythologization of TE. If teachers do not know what characteristics exemplify technology and TE, then TE is to them a mystery (Daugherty and Wicklein, 2010:2). If teachers' understanding of certain concepts about TE is skewed or incorrect, then myths about what TE really is will develop. It is these mysteries and myths that if not studied and cleared will culminated in the thought that TE should be deleted from the curriculum as was recommended by the review committee report (Chisholm, 2000:125).

5. Investigation of How Teachers Understand Technology and TE

An investigation of how teachers understand the concepts technology and TE and how they think this (TE) should be implemented in the curriculum is correlated with the issues of demythologizing them. We asked the following questions:

- What are teachers' understanding of the concept "technology"?
- What are teachers' understanding of the concept "TE"?
- What are teachers' perceptions about the implementation of TE in the curriculum?

6. Methods

Both quantitative and qualitative methods were used by exploring teachers' understanding of what technology and TE entails, together with how TE should be implemented in the curriculum to bring the success of the implementation of the very TE. The quantitative method tries to capture the perception of teachers through a depiction of their views by constructing aggregate percentages through interviews. Because reality is not necessarily a matter of numbers only, it was advisable to combine both the quantitative and the qualitative methods. The essence of this can be corroborated by Albert Einstein's deliberation that "Not everything that counts can be counted and not everything that can be counted counts" (Kevin Harris, 1995:3). Thus the in-depth or thick descriptions characteristic of the qualitative methods (Mouton and Marais, 1993:67) would be used to complement the numerical quantitative outcomes of the study. This is called triangulation and adds further validity to the research.

7. Sampling

An exploratory study was conducted on a sample of 300 TE teachers. The purpose of the exploratory study was to establish teachers' understanding of TE with respect to the three themes selected for this study, that is, the concept "Technology", the concept "TE" and "TE in the curriculum." This was a quantitative study in which a questionnaire was used to collect the data. The 300 teacher respondents were selected randomly from three provinces, that is, Mpumalanga (180), Limpopo (118) and Gauteng (2).

It must be indicated here that the Gauteng province was originally not part of this study. How the two responses from Gauteng (or as indicated on the questionnaire) became part of the respondents is unclear. The researcher had a choice to reject these responses or to take them in as part of the sample. Taking them as part of the sample meant that the Gauteng province formed part of the study.

Based on the outcomes of the exploratory study, a data collection schedule for the qualitative study was constructed. The qualitative study was conducted on 22 teachers randomly selected from the three provinces as stipulated above. However, it must be acknowledged that the random selection of the qualitative sample may have included teachers who were not part of the original 300 quantitative study sample. Based on the purpose of the study, this deviation is not significant and as such does not pose any validity threat.

The purpose of the qualitative study was to reinforce or refute the findings of the quantitative studies. Because of the inherent attitudinal disposition of the concepts "knowledge" and "understanding" as used in this study, using only the quantitative methodology would jeopardize the validity of the study and would make the outcomes of the study to be untrustworthy. Besides, collection of conceptual data such as used in this study lacks scientific theory (Pudi, 1999:5). According to Vockell and Asher (1995:192), "the qualitative research admits that the analysis of human activities is by necessity largely subjective and attempts to make the subjective analysis of the human behaviour as unbiased as possible." That is why this methodology was chosen to complement the quantitative methodology used.

8. Data collection instruments and techniques

Consistent with the data collection methodologies for this study, a questionnaire and interviews were used. The questionnaire consisted of nine statements. These statements were devised to elicit information on teachers' understanding of technology and TE. The statements were derived from the three main themes of this study. These themes are part of the ongoing problems about TE in the curriculum and have been devised on the basis of the department of education circulars; the review report and other literatures concerned with problems in the implementation of TE. These statements were based on a 5-point Likert scale. The purpose of this was to elicit responses which indicated teachers' knowledge and understanding of Technology, TE and related concepts on a sliding scale. Nachmias and Nachmias (1992:239), point out that the purpose of the questionnaire is to translate the research objectives into specific questions or statements.

The essence of the 5-point Likert scale as used in this study was to increase the variations in the possible scores by coding from "strongly agree" to "strongly disagree" instead of merely "agree" or "disagree" (Daugherty and Wicklein, 2010; Bailey, 1987:346). The responses were rated so that a higher score for a particular statement indicated stronger agreement with the knowledge or understanding. The situation can be depicted as:

Response	**Rating**
Strongly agree	5
Agree	4
Uncertain	3
Disagree	2
Strongly disagree	1

The middle response, coded 3 depicts a state of indecision or do not know.

For interviews, a schedule was constructed. This consisted of sub-themes as shown in Table 1 together with a number of points emanating from the questionnaire which needed further clarification through interviews. The interviews were in fact a way to seek for better consensus of the findings, thereby increasing the validity of the results.

9. Analysis and interpretation

Findings and interpretation were done according to the three main categories or themes, that is, the concept "Technology"; the concept "Technology Education" and "Technology Education in the curriculum" as shown in Table 1.

The empirical understanding of the three themes was investigated through both qualitative and quantitative methods. Quantitative responses to each of the statements were tabulated for interpretation. The interpretation took into account both the quantitative and the qualitative responses together with the theoretical foundations of the three concepts as discussed in the opening part of this study. For the sake of the quantitative analysis and interpretation of the results, the 5-point Likert scale was reduced to the 3-point Likert scale.

Table 1: Schedule of themes and sub-themes

Theme	Sub-theme	Relevant Statements
The concept "Technology"	Technology as machinery Technology as scientific knowledge Technology as synonymous with computers Technology as a social process	*"Technology is all about machinery"* *"Only learners who have previously taken science at school should enroll for a course in TE"* *"When I think of technology, I mostly think of computers"* *"Technology is a social process"*
The concept "TE"	TE as a male domain	*"Only boys should take TE at school"* *"Female teachers should not be involved in facilitating or teaching TE"*
"TE in the Curriculum"	Reinstatement of TE in the Curriculum Integration across learning areas TE as problem solving	*"Although it was re-instated, the proposal to drop TE as a learning area from the curriculum by the review committee on C2005 was appropriate"* *"Integration across learning areas is easily achieved through TE"* *"Technology Education is about problem solving"*

(i) Responses to the statement: *"Technology is all about machinery"*

Agree: 28.7 **Disagree: 65.4** **Uncertain: 5.9**

The conception that technology is machinery is an old stigma. This is confirmed by Naughton (1986:2) that "equating technology with machinery, sometimes referred to as hardware, is common nowadays." The 28.7% responses that agree with the statement that technology is all about machinery confirms this belief. According to Naughton (1986:2), the equation of technology with machinery may be valid in the sense that it represents common usage, such as referring to a house as a machine for living in. One of the respondents from the group that was selected for interviews wanted to know what the interviews will contain so that he can bring along what he called "my little calculating machine." Instead of seeing the wrong, the rest of the group was put in awe as demonstrated by the remark:

> *"We hope that this is not about calculations"*

While this was a clear demonstration of the phobia for figures, reference to the calculator as machinery was acceptable to the group. What emanated from the discussions that followed was that the department of education was blamed for lack of proper introduction of TE. The respondents said that they have to guess many things.

The quantitative demonstration that technology is not all about machinery (65.4% responses) was reinforced by the verbal responses that it is about artifacts, and everything that is being done with the hands including machinery. This conforms with Grimsley (1999:1), that "technology allows students to reinforce, apply and transfer their academic knowledge and skills to a variety of interesting and relevant activities, problems and settings." Technology is thus not just about machinery. One of the interviewees rightfully remarked:

> *"Machinery is just but one aspect in the TE curriculum, just like pneumatics or electricity".*

This is insightful, taking into consideration that the policy document (DoE, 2002:36) does not stipulate topics for the curriculum but merely depicts the curriculum in terms of the three Learning Outcomes. The three Learning Outcomes are broad statements from which the teacher has to deduce the topics and the activities that he or she needs to teach to achieve the desired Learning Outcomes. However, the broadness of the statements did not prohibit the respondents from exhibiting the understanding that machinery is part of TE just like pneumatics or electricity as depicted by the remarks above.

(ii) Responses to the statement: *"Only learners who have previously taken Science at school should enroll for a course in TE"*

Agree: 7.9 **Disagree: 86.4** **Uncertain: 5.7**

The majority of the respondents (86.4%) agree that Science is not a prerequisite for TE. However, 13.6% are either uncertain or believe that science is a prerequisite to take TE at school. That the majority of respondents indicated that science is not a prerequisite to take TE is a positive step towards understanding the purpose of TE in the curriculum (Ankiewicz, 1995:3). The contrary belief would have resulted in a lot of learners being deprived or excluded from taking TE. This would have gone against the vision of the Department of Education (DoE, 1997; Ort Step Institute, 1995) to make "every boy and girl to be technologically literate by the year 2005."

The 13.6% responses may be due to the stigmatization of technology as an off-spring of science. A comment was made during the interviews that learners who were accepted for TE without the background of Science do not do well.

(iii) Responses to the statement: *"When I think of technology, I mostly think of computers"*

Agree: 54.4 **Disagree: 42.1** **Uncertain: 3.5**

The majority of the respondents (54.4%) agree with the statement. This confirms that most respondents still think of computers as technology.

Most respondents still perceive computers as exclusively technology as was demonstrated during the interviews. While computers alone may not be technology, they are in fact part of technology and TE (Pudi, 2002:45-46). Technology extends beyond only computers, it can be an industrial process or a social process; it is means and ways to solve problems and to achieve needs and wants (DoE, 1996:1, and 2002:4). According to Naughton (1986:3), technology also involves organization, thinking and doing. It is thus imperative to define the parameters of technology and TE so as to off-set amongst others, the misthought or understanding that technology and TE are only about computers.

According to Solomon (1993:55), "Many young people simply equate technology with computers." This is corroborated by Moodic (2009:1) as alluded in the previous discussions above that in the public perception, technology is the same as computers. During the interviews some of the respondents mentioned that despite their try to see TE in its true sense, their colleagues keep asking them to repair or help them with some computer problems clearly on the belief that TE is about computers.

Computers play a major role in the TE approach since they are an integral part of modern technology. According to Pullias (1992:3), the emphasis in TE is on learning how a computer is used as a tool more than learning "about" computers. The concern in TE is not so much the proficiency in a certain skill, as the learner's understanding of concepts, relationships, problem solving, analysis and evaluation (technological process) because technology education is a series of related projects.

A significant number of respondents who either disagree or are uncertain, either do not associate technology with computers or are unaware that Technology and TE are not only about computers.

(iv) Responses to the statement: *"Technology is a social process"*

Agree: 15.2 **Disagree: 76.6** **Uncertain: 8.2**

The majority of the respondents (76.6%) disagree with the statement. This is contrary to the ideals of Learning Outcome 3 for TE. Learning Outcome 3 stipulates that "the learner should be able to demonstrate an understanding of the interrelationships between science, technology and the environment. Technology as a social process conforms to this stipulation (Learning Outcome 3). Technology is to be found in all spheres of life (Pudi, 2002:87). This is catered for by the stipulation 'society and the environment' in Learning Outcome 3 as described above.

Technology can be a scientific process or it can be "a spontaneous process as in crafts, like pottery, where there is no form of social organization" (Cross, Naughton and Walker, 1986:9). Based on this deliberation, it is clear that technology is a social process. The understanding of TE as a social process is crucial because in TE learners make artifacts and solve problems in their own environments.

The basis for the misthought that technology is not a social process was clarified during the interviews. In general, the interviewees or respondents believed that technology and hence TE is not a social process as it takes place in the school and not in the community or 'society'. Clearly the school was not seen as part of the community and the community was not seen as a place or the environment where technological problems are applied. While TE takes place within the school, its application and benefits are to the society (Dryden and Vos, 2001).

(v) The responses to the statement: *"Only boys should take Technology Education at school"*

Agree: 95.6 **Disagree: 2.6** **Uncertain: 1.7**

An overwhelming majority (95.6%) believe that TE is good for boys and by deduction, not suitable for girls. Perhaps this stems from the myth that technology has more to do with hands and that girls cannot do manual work. Through Learning Outcome 1, the Department of Education (DoE, 2002:6) describes technology as a practical solution-oriented learning. Life's problems confront all men, women, boys and girls. Thus technology as a practical solution-oriented learning is necessary for all – irrespective of gender.

An advert in the 'Sunday Times Magazine' depicts the myth that technological careers are for boys while females and girls are associated with rearing children and playing with dolls (Anon, 1996:15). This is a very old myth. Positive responses towards the statement that 'only boys should take TE at school' reinforce this myth.

The stigma that TE is a male domain is derived from an association with industrial arts and technical subjects. This belief overlooks the fact that technology and consequently TE are to be found in all spheres of life (a social process). Furthermore, females are to be found in all spheres of life where their male counterparts are and not only in industry.

According to Zuga (1999:1-3), "women and girls often perceive the subject of technology education as a male domain, especially after they have taken a course in it." These findings have been proved to be true by the respondents in this study. The question arises as to what is it in TE that makes women and girls, especially those who have been exposed to TE to see this subject as a male domain. There has been a thought that cultural roles may be playing a part in this. However, in the TE curriculum there are projects such as sewing and cooking which are perceived to be the role of women. Such projects, one wants to argue, are those projects that should sway women and girls' perceptions from thinking that TE is a male domain. If anything, the argument can be extended that men and boys can, to the contrary, see TE as a female domain because of the inclusion of those projects.

While technology prioritizes the effective use of technological products and systems, it is not gender biased (Grimsley, 1999:1; Van Rensburg *et al.*, 1996:2; Treagust and Mather, 1990:53; Hansen, 1999:244-296). Mitter (1993), reports that women take part in many technological endeavours including even in the military where some are pilots and aeroplane technicians and many men have been employed as chefs and nannies.

Perhaps the earlier deliberation that girls do not do well in TE as was said in the interviews has a contribution to the belief that only boys should take TE at school (refer to the statement *"only learners who have previously taken science at school should enroll for a course in TE")*

(vi) The responses to the statement: *"Female teachers should not be involved in facilitating (teaching) TE"*

Agree: 9.6 **Disagree: 90.0** **Uncertain: 0.4**

An overwhelming majority (90%) disagrees with the statement. Disagreement in this case implies agreement with the assertion that "Female teachers **should** be involved in teaching TE." Although in my opinion this is correct and relevant, it is however a clear contradiction with the findings in the previous question about boys and girls where 95.6% of the respondents believe that only boys should take TE at school. An argument is then raised: If female teachers are to teach TE as demonstrated by the responses, then girls should also take TE at school. In short, if teaching TE is for females, then girls must take TE at school. TE has no gender bias, whether this be towards girls and female teachers or towards male teachers and boys (Pudi, 2002:100).

Based on the assertion that *only boys should take TE at school,* and consequently seeing TE as a male domain, it is hard to understand why females should be involved in teaching TE. A further question may be raised: If girls do not do well in TE (as described in the interviews), how then is it possible that female teachers can do well as TE teachers? The deliberation that *"Technology is a social process"* that involves both male and females could be used as proof that both male and female teachers deserves to teach TE and that boys and girls can and should be given equal opportunities to learn TE. But, according to the discussions on this topic, the respondents do not see Technology as a social process. This would have settled the argument.

The above responses are very interesting when one takes into consideration that 56% of the respondents for this study were female TE teachers. In this technocratic age of women liberation where women in particular are sensitized to any forms of discrimination, it is unbelievable that such a glaring discrimination practice as this one can be overlooked – even by women themselves. Thus to discriminate against women teaching TE is to refuse them their constitutional right to participate freely in all human endeavours in society. According to Pudi (2002:101), there is nothing collectively towards the teaching of TE that makes women to be inferior to men and nothing that makes men better suited for TE above women.

The sum of the responses 9.6 (Agree) and 0.4 (Uncertain) equals 10.0 which is insignificant in this case.

(vii) Responses to the statement: *"Although it was re-instated, the proposal to drop Technology Education as a learning area from the curriculum by the review committee on Curriculum 2005, was appropriate"*

Agree: 15.8 **Disagree: 65.6** **Uncertain: 18.5**

The majority of the respondents (65.6%) disagree with the statement. This in essence means that they are not happy with the thought of dropping TE from the curriculum. This unhappiness was demonstrated beyond doubt during the interviews. This is interesting because the recommendation to put TE to extinction was as a result of teachers' dissatisfaction with it as captured in the curriculum review report (Chisholm, 2000:4). Posing this contradiction to the interviewees revealed yet another mystery. All the twenty-two interviewees were adamant that they were not part of the sample that was selected for the C2005 Review. There was an indication that the sample for C2005 review may have been composed of respondents from the more urban provinces of which is not the case with the majority of the sample for this study. As a result, an impression was created that their responses would have been in favour of retaining TE in the curriculum if they were part of the sample for the review. Based on the reality that within this very group (which was allegedly omitted from the C2005 review) there is 15.8% responses which agrees with the statement that it was appropriate to think of deleting TE from the curriculum, it is hard to believe that their inclusion in the review sample would have made any significant difference to support the motion to retain TE in the curriculum. My reluctance is strengthened by the 18.5% who did not commit themselves.

However, it is illogical to expect that TE teachers would support any move to have it (TE) removed from the curriculum. Being involved with a subject has emotional attachments. It is these emotions that are nuances when it comes to finding the truth about a situation. As a TE teacher himself, the researcher is also exposed to these emotions.

Welcoming the re-instatement of TE in the curriculum is thus logical. However, it is advisable to undertake an ongoing review of the reasons that contributed to the thought of removing TE from the curriculum. Reasons that led to the re-instatement of the TE learning area should also be revealed and be made part of the ongoing examination and improvement of TE.

(viii) Responses to the statement: *"Integration across learning areas can be achieved through Technology Education"*

Agree: 11.8 **Disagree: 73.8** **Uncertain: 14.7**

The majority of the respondents (73.8) disagree with the statement. This points to lack of understanding of what technology really is. The respondents perceive TE as something completely new, maybe because of its newness as a learning area in the curriculum or lack of knowing and understanding its purpose and parameters. The very nature of TE allows for integration across learning areas and across the curriculum (Pullias, 1992:1; Lewis, 1999:8). The 11.8% who agree with the statement reflect adherence to the curriculum policy expectations of integration (DoE, 2002:5). The 14.7% who are uncertain may have derived their uncertainty from the lack of proper understanding of TE and its rationale in the curriculum.

Integration of TE across the curriculum should not be confused with Ankiewicz's deliberation (Ankiewicz, 1995:1), that "the subject of Technology Education should be viewed as a subject with its own epistemology, philosophy, aims, identity, structure, method of inquiry, curriculum, didactics and opportunities for the promotion of problem solving and other higher order cognitive skills." TE has indeed a right as a subject with its own identity, but concepts used in TE can be integrated in other learning areas.

If schooling is to have the desired meaning for learners, then the various elements of the curricula must cohere. Lessons learnt in one subject area must be amplified in others (Lewis, 1999:9). To take its place squarely in the curriculum, TE must establish itself not just in its own right, but critically in relation to other subjects (Lewis, 1999:9). Contrary to the good intentions of Lewis, the tendency is that TE may be both a mystery and a myth. Admittedly it may be difficult to identify technology and TE across the learning areas and in the learning areas themselves due lack of proper understanding of what technology and TE really are.

The response that TE does contribute to the integration across learning areas corroborates with the earlier responses where respondents did not see Technology as a social process which exist in the "society and in the environment" (Learning Outcome 3).

Some of the comments made during the interviews alluded to the difficulty of TE integrating the curriculum for various reasons:

> "The curriculum was there before TE. I do not see how TE can integrate the curriculum when there is still a struggle to retain it in the curriculum."

> "Instead of integrating, TE must first succeed in establishing itself as a bona-fide part of the curriculum."

(ix) Responses to the statement: *"Technology Education is about problem solving"*

Agree: 25.4 **Disagree: 74.6** **Uncertain: 0.0**

The majority (74.6%) disagrees with the statement whilst 25.4% are certain that TE is about problem solving. The negative response to the statement that TE is about problem solving was found to emanate from the respondents' narrow thinking that TE is about making the artifacts. There was a debate as to whether making artifacts is problem solving or not. The thinking that TE is about making artifacts may have its roots from the thinking that TE is about manual work, which could result in it (TE) not being seen as suitable for girls. The argument that girls also have hands and can also do manual work cannot be repeated here.

Technology is about problem solving in practical relevant problems that are worth solving (Middleton, 1998:1; Potgieter, 1998:13). Robert Sternberg (Pudi, 2002:169) emphasizes identification of a problem as the first step in problem solving. If TE is about problem solving, the problems must first be identified. Making the artifact is a way of solving the identified problem.

According to the discussion of Learning Outcome 1 (Technological Processes and Skills) in (DoE, 2002:6), TE can be seen as a practical, solution-oriented learning. This is a confirmation of Ankiewicz's suggestion (Ankiewicz, 1995:3), that the TE curriculum should provide for opportunities for the promotion of problem solving and other higher order cognitive skills. Chetty (1998:1) sees the beauty of how to solve problems as being the pivotal point in TE for C2005. This point is stressed by Middleton (1998:1), when he advocates that problem solving is central to the practice of TE.

A comment was made that *"TE is about making artifacts and products"*. Based on this comment, one wonders whether making artifacts is not solving a problem.

10. Findings and discussions

The findings for this study will be based on the three research questions together with the sub-themes as shown in Table 1. These will be the findings from both the quantitative and qualitative outcomes of this study.

With respect to the theme *"Technology as machinery"*, it is clear that the understanding of the term "technology "is still elusive. Although most respondents indicated that Technology is not all about machines, there is still a significant number that associated technology with machinery. This was found to be a problem of how technology was introduced. The stigma that technology is machinery has been demonstrated by even referring to a calculator as a machine. As far as seeing technology as *"scientific knowledge"* is concerned, the findings are that there is a very strong belief that technology is part of science. Consequently, respondents still have a belief that learners should first have a scientific background (passed science) in order to take TE as a subject. For TE to be recognized as a subject in its own right, the respondents believe that it should not form common areas with other subjects. In essence this is like drawing walls between parts of education. According to Pullias (1992:3), this is like "dividing education into different cubicles or departments and putting TE into one." Reference to *"Technology as synonymous"* with computers is strangely still in existence. There is however acknowledgement that the respondents are somehow forced to think of technology as the same with computers because of the attitudes of their colleagues who have an expectation that, because they are TE teachers then they are also computer experts. *"Technology as a social process"* is understood narrowly to mean that TE only takes place in school. The school is not seen as part of the community or societies. With respect to the theme *"TE as a male domain,"* there is a general belief that TE is only for boys. This is despite the fact that a large number of respondents (56%) were females. Perhaps the response was compelled by the remark that "girls do not do well in TE". However, this was well contradicted by the overall disagreement to the statement that "Female teachers should not be involved in teaching TE."

As far as *"Reinstatement of TE in the curriculum"* is concerned, there is still sad remnants of the intentions to have it (TE) scrapped from the curriculum. Some respondents, although conceding to the facts that TE has added to their workload, are adamant that they did not form part of the sample that was consulted for the Review committee and feel that if they had been consulted, would have fought to retain TE. The problem about TE from the respondents' perspective is that there has not been any teacher training and that the department of education does not offer support, especially towards TE. They hardly have tools and material to conduct TE well. There is a belief (as confirmed in the interviews) that TE seems like a misfit in the curriculum, because its parameters are not well defined. Despite this, there is still belief that TE has the power to integrate the curriculum ("Integration across learning areas") but that that power has not been exploited.

In as far as the sub-theme of "TE as problem solving" is concerned, there is a belief that TE is about making artifacts as if making artifacts is not problem solving.

Based on the preceding discussions, it is clear that teachers' understanding of the concept "technology" is not what it should be. The same is true for the concept "TE". These may have detrimental effects towards their teaching of TE. These might compromise the achievements of the Learning Outcomes for TE. By their very responses, teachers have indicated lack of confidence in themselves in as far as the implementation of TE in the curriculum is concerned.

11. Conclusions and Recommendations

Teachers' understanding of the terms Technology and TE is of paramount importance for the proper implementation of this learning area. Clear understanding of these terms will alleviate the propagation of myths and stigmas which would have a negative impact in the implementation of TE.

According to the findings, the respondents do not have a clear vision of what TE is. As a result, they will fail to take advantage of its true nature and most likely see TE as just an academic exercise that does not go beyond the four walls of the classroom or the workshop. Learning Outcome 3 which advocates for the interrelationship between society and the environment will not be achieved for as long as teachers lack this understanding. If the stipulations of Learning Outcome 3 seem to be a mystery, educational authorities should encourage demystification through advocacy or rewriting these in simpler form.

In as far as lack of better introduction and curriculum support is concerned; it is recommended that TE learning area support groups should be formed. Here teachers should discuss the pros and cons of TE from policy to classroom realities. There should also be conferences organized where teachers will be invited as guests. These conferences should focus on TE and the conference proceedings thereof should be captured in hard copy and electronically and distributed to the teachers.

To support TE, curriculum implementers must be appointed to oversee on a regular basis what is taking place in the TE classes and to offer support. Time for TE should also be increased in the time table to allow TE teachers to make up for the lost ground caused by delays as a result of doubting the legitimacy of TE in the curriculum.

The legitimacy of TE in the curriculum is based on its purpose and rationale to be in the curriculum itself. This (purpose) is linked to learners and teachers' understanding of what technology and TE are. This study has revealed that there is still misunderstanding and misrepresentations of technology and TE. This has a number of features that are common with other studies around the world in that technology is often linked to science and is often seen narrowly as information technology only. This has led to the false belief that technology and consequently TE is thought of as computers. It is recommended that advocacy campaigns should be emphasized where technological terminology, together with its rationale and framework should be clarified.

In as far as integration across the curriculum is concerned, TE teachers must be compelled to indicate integration with every Learning Programme that they develop. Integration across learning areas makes work easier in that the integrated aspect is learned more than once but has the added advantage of being seen from different perspectives, for example, from a technological perspective, a scientific perspective or mathematical perspective depending on which subject the integration is made with.

Integration may also be seen as a means to desmystify and demythologize technology and TE, by relating it to other learning areas within the curriculum. If lessons learnt in TE cannot be applied in other learning areas and vice-versa, then the different part of the curriculum would be seen not to cohere. Through integration, TE specifically, will be made a necessary part of the curriculum. The integration of TE with other LA's will make it difficult for it (TE) to be the casualty of curriculum pruning as it has already been attempted.

Demystification and demythologization of technology and TE will enable teachers to take advantage the *"good things"* of technology and TE and spread these to their learners. The *"good things"* in the study of TE are according to Grimsley (1999:1) "reinforcement, application, and transfer of students' academic knowledge and skills to a variety of interesting and relevant activities, problems and settings" which (we add) include problems within the society and the environment.

References

Ankiewicz, P. (1995), 'The Planning of Technology Education for South African schools', in *International Journal of Technology and Design Education,* 5(3): 1-10.

Anon, (1996), 'South African pupils are the dumbest in the world', in *Sunday Times magazine*, December 1996.

Bailey, K.D. (1987), *Methods of social research,* 3rd edition, New York: The Free Press.

Black, P.J. (1998), 'An international overview of curricular approaches, and models in technology education', in *The Journal of Technology Studies,* 24(1):24-30.

Brown, R.A. and Brown, J.W. (2009), 'What is Technology Education? A Review of the "Official Curriculum", in *The Clearing House,* 83*:* 49-53.

Chetty, P. (1998), 'Technology in the new curriculum', in *Spectrum* 36(1): 2-3.

Chisholm, L. (2000), *A South African Curriculum for the twenty first century: Report of the Review Committee on Curriculum 2005.* Available at: http://education.pwv.gov.za/Policies_Reports/Reports_2000/2005.htm

Cross, N., Naughton, J. and Walker, D. (1986), 'Design method and scientific method', In: Cross, A and McCormick, R., (ed.), *Technology in schools: A reader*, New York. Open University Press (pages 18 –33).

Daugherty, M.K. and Wicklein, R.C. (2010), 'Mathematics, Science and Technology Teachers' Perceptions of Technology Education'. Available at: http://scholar.lib.vt.edu/ejournals/JTE/v4n2/daugherty.jte-v4n2.html.

Department of Education (DoE) - South Africa (1996), *The Technology Learning Area Committee Workshop Report*, 20 September, Pretoria: Government Printers.

Department of Education (DoE) - South Africa (1997), *Lifelong learning for the 21st century.* Pretoria: Government Printers.

Department of Education (DoE) - South Africa (2000), *Comments on Report of Curriculum 2005 Review Committee*, 19 June, Pretoria: Government Printers.

Department of Education (DoE) - South Africa (2002), *Revised National Curriculum Statement (RNCS) Grades R-9 (Schools) – Technology*, Pretoria: Government Printers.

Department of Education (DoE) - South Africa (2009), *Report of the Task Team for the Review of the Implementation of the National Curriculum Statement*, Final Report, October, Pretoria: Government Printers.

Dryden, G. and Vos, J. (2001), *The Learning revolution: To change the way the world learns*, New Zealand: Network Education Press Ltd.

Dugger , W. and Yung, J.E. (1995), *Technology Education Today*, Bloomington: Phi Delta Kappa Educational Foundation.

Gaganakis, M. (1999), 'Failure and success: a critical literature review of gendered performance of black adolescent girls in South African schools', in *South African Journal of Education,* 19(3): 147 -153.

Grimsley, R. (1999), *Technology Education Program Evaluation*: Texas Education Agency. Version 1: July 1999. Available at: www.tea.state.tx.us.

Hamm, C.M. (1993), *Philosophical Issues in Education: An Introduction*, New York: The Falmer Press.

Hansen, R. (1999), 'The Technology Workshop at Oundle', in *International Journal of Technology and Design Education,* 9(3): 293-303.

Harris, K. (1995), *Collected Quotes from Albert Einstein*, pages 1 -3. Available at: http://rescomp.stanford.edu/~cheshire/EinsteinQuotes.html (Accessed: 18 December 2009).

Hodgkin, R.A. (1990), 'Techne, Technology & Inventiveness', in *Oxford Review of Education*, 16(2): 207-217.

Jeevenantham, L.S. (1999), 'Towards C2005: Proposals for curriculum change', in *South African Journal of Education,* 19(1): 49-66.

Kramer, D, (1994), 'Technology Education: A hit or a myth?' *Paper presented to board of directors at PROTEC*, Johannesburg: PROTEC.

Lewis, T. (1999), 'Research in Technology Education – Some Areas of Need', in *Journal of Technology Education,* 10(2): 10-13.

McCormick, R. (1997), Conceptual and Procedural Knowledge. *International Journal of Technology & Design Education,* 7(1-2):141 – 160.

Middleton, P. (1998), *Protec Technology Education Curriculum*, Manzini: Macmillan Boleswa Publishers.

Mitter, S.S. (1993), *Dignity and daily bread new forms of economic organizing among poor women in the Third World and the First*, New York: Rowbotham Routledge.

Moghaddam, F.M. (1998), *Social Psychology: Exploring Universality Across Cultures*. New York: W.H. Freeman & Company.

Moodie, P. (2009), 'A response to the draft Report of the Task Team for the Review of the Implementation of the National Curriculum Statement', September 2009, Setlhare Science Curriculum Trust, Johannesburg.

Mouton, J. and Marais, H.C. (1993), *Basic concepts in the methodology of social sciences,* Pretoria: Human Science Research Council (HSRC).

Nachmias, C.F. and Nachmias D. (1992), *Research methods in the social sciences*, Kent : Hodder Stoughton.

Naughton, J. (1986), 'What is 'technology' anyway', In: Cross, A. & McCormick, R. (eds.) *Technology in Schools*, Philadelphia: Open University Press, 1-9.

O'Donovan, E. (2009), 'Demystifying emerging technologies'. Available at: http://web.ebscohost.com/ehost/delivery?vid=6&hid=7&sid=e51f7790-6021-474a-bc3f-e...: 1 – 5 (Accessed: 16 February 2010).

Ort-Step Institute (1995), *Proposal for primary school technology curriculum*, Midrand, Johannesburg: ORT - STEP Institute.

Poon, A. (1994), *Tourism, Technology and Competitive Strategies*, Wallingford: CAB International.

Potgieter, C. (1998), *Technology A: Only study guide for FDETE1-A*, Pretoria: UNISA.

Pudi, T.I. (1999), 'Critical and creative thinking skills in Technology Education', *M.Ed Dissertation*, University of Johannesburg, South Africa.

Pudi, T.I. (2002), 'Teacher attitudes towards the implementation of the learning area Technology', *D.Ed Thesis*, Pretoria: UNISA.

Pudi, T.I. (2009), 'The product or the process? Is Technology Education all about "Making"?' *Unpublished Paper*.

Pullias, D. (1992), 'What is Technology Education?' in *The Technology Teacher,* 51(4): 3-4.

Rudinow, J. and Barry, V.E. (1984), *Invitation to critical thinking*, 2nd Edition, Fort Worth: Holt, Rinehart & Winston.

Shield, G. (1996), 'Towards defining and evaluating concepts and strategies for process within technology education', *PATT Conference*, 17-22 April 1997, The Hague.

Solomon, J. (1993), *Teaching Science, Technology and Society*, Buckingham University Press.

Steinem, G. (2008), 'Thoughts on Quotes'. Available at: http://thoughtsonquotes.blogspot.com/2008/02/learning-is-easy-unlearning-is-hard.html.

Texas Education Agency (2010), Available at: www.tea.tx.us (Accessed: 25 March 2010).

Treagust, D.F. and Mather, S.H. (1990), 'One school's approach to Technology Education: Integration across the curriculum', in *The Australian Science Teachers Journal,* 36(3): 50-60.

Van Rensburg, S.J., Myburgh C.P.H. and Ankiewicz, P. (1996), 'Curriculum development for Technology Education in South Africa: Gender issues', *Paper presented at the GASAT 8 Conference*, Ahmedabad, India, 5-10 January.

Vockell, E.L. and Asher, J.W. (1995), *Educational Research*, Englewood Cliffs: Prentice Hall.

Volti, R. (1988), *Society and Technological Change*, New York: St. Martin's Press.

Warner, S.A., Elrli, L.A., Johnson, C.W. and Greiner, S.W. (2007), 'Identifying the Paradigm of Design Faculty in Undergraduate Technology Teacher Education in the United States', in *Journal of Technology Education,* 18(2): 68–89.

Williams, P.J. (1992), 'Design: an appropriate Technology Education methodology in less-developed countries', in *Journal of Epsilon Pi Tau,* 18(1): 53–60.

Zuga, K.F. (1999), 'Addressing Women's Ways of Knowing to Improve the Technology Education Environment for All Students', in *Journal of Technology Education,* 10(2):1–15.

Chapter 8

Local Universities as Engines for Innovation, Take-Off & Expansion of Regional Development in Southern Economies

Ahmed Driouchi & Nada Zouag

1. Introduction

A series of studies and publications still refer to the increasing quantitative and qualitative diversities of needs of the population in developing economies and mainly in Africa (De Pee et al., 2010; McTavish et al., 2010; Quisumbing, 2010; Kebede et al., 2010; Anderson et al., 2010 and Beneria, 2010). According to the mentioned studies, these needs concern almost every component related to health, education, employment, income and infrastructure. Furthermore, as these needs are expressed locally by households in both rural and urban areas, their satisfaction cannot be only from local supplies. But, responses and supplies can sometimes be locally provided with possibilities of the development of further trade and exchange with other localities in the same country or outside. But potential suppliers may need further knowledge and also inputs from research and innovations ensuring thus, the competitiveness of the new products and services. Several authors have looked at different dimensions of the links between R&D and local supplies. These include Rothwell (1977), Pavitt (1984), Reinert (2006), Lundvall (2007), Amsden (1989), Freeman (2004), Mazzoleni and Nelson (2005), Peres (2006), Stiglitz (2001) and Reinert (2007).

The major objective of this paper is to show that local and regional universities can play an important role in the process of providing knowledge to contribute to the supply of the desired and growing needs of local populations. This is shown to be related to innovation and the generation and development of integrated technological and socio-economic platforms related to the valuation of new and specific local and regional niches and opportunities.

This paper will focus on the impact of new localized universities on the territorial and general economic growth and development. It attempts to answer question such as: Can these universities along with tertiary education be engines leading to innovation take-off and local development? If this is tested to be relevant, then what promises and prospects exist for most developing economies?

The paper provides a literature review in the following section about the role played by universities in local development in different countries and regions. The overall assumption and hypotheses to be tested is that there are three groups of countries and that these groups perform differently, given the way they combine new and old sources of growth and development. This is tested in the second part of the paper showing that developed countries have a better pattern of local economic development than developing countries. Besides that, transition economies have an improved pattern, compared to developing countries. Following the empirical methods pursued to test the hypotheses; the practical implications of the results are then provided and discussed.

2. Literature Review

Local development in different countries and mainly in developing economies has been traditionally based on the valuation of traditional business ideas and projects with preservation and control of the gained positions. These businesses have been generally benefiting from autarkic means of preservation and protection, including familiar transmission of tacit knowledge within the traditional structures that monitor the process. These types of models are becoming progressively obsolete and local economic development can now be driven by new ways of thinking where advanced knowledge occupies a dominant position. These new models have been already experienced in developed economies where most of time, current knowledge has been replacing traditional ways of doing business. The new drivers for developing economies are to be based on specific engines that are primarily related to the contributions of education and the training offered by the local universities. It has been observed that university activities in regions play an important role in promoting local development. This is a new approach where knowledge is the key factor for future economic competitiveness and progress. Therefore, universities and tertiary education systems explicitly play the major role.

Shaffer and Wright (2010) explain that universities and higher education systems can act as drivers of economic development and community improvement through using their research force and developing innovative ideas to reinforce the competitiveness of the economy and use innovations commercially and socially. They can also provide knowledge-based services to businesses and employers, educational, social and cultural revival of local communities and education for the success of people as individuals and groups.

Hill (2006) offers some incentives as to the extent to which university research, especially faculty innovative output, supports local economic development. The author also points to the contributions made to the local economy through research universities and their graduate programs. The reason of this success is due to the ability to codify knowledge derived from research efforts, which can reach everyone not only the closest population, while the transfer of tacit knowledge is surrounded by obstacle. Universities also influence local economic development through retaining graduates in the vicinity of their schools especially if the university is located in a large urban area. The skills and field of study of those graduates are, for example, dependent on labs and qualified science and engineering training (Hill, 2006). University research is also found to influence local economy development through promoting corporate research and development (R&D) activity. Besides that, results from a number of econometric studies showed, through the use of variables measuring economic activity and others for the research university, a positive statistically significant relationship in many cases but the strength of the relationship usually seems weak (Hill, 2006).

Ji et al. (2010) found that local colleges play an important effect on Chinese regional scientific and technological innovation. They are considered a major node in Chinese regional innovation system through thinking and understanding activities, production and research cooperation and joint scientific and technological research among others. Frenette (2009) found that the creation of new universities, in cities where there were none, did indeed increase university attendance among the local youth at the expense of colleges in most cities. However, the short run effect of new university establishment is the considerable rise of the probability of moving out of a given city for both men and women. Men experience increased employment and women get an increase in business services and public sector employment (Frenette, 2009).

Westnes et al. (2009) worked on a comparative study of the role played by local universities and public research organizations in the promotion of local capabilities for innovation. This latter study was performed on two gateways to the North Sea oil and gas province which are the Stavanger region on the southwest coast of Norway and the Aberdeen region in northeast Scotland. The authors found that the contribution of education and research organizations to regional development of both oil regions was made differently but there is no proof affirming that one is better than the other in developing competitive and powerful regions. In the Stavanger region (Norway), universities developed relevant technological capabilities based on coordination and collaboration. The Aberdeen universities (Scotland) developed ties to industry through the achievements of individual academics in various disciplines (Westnes et al., 2009).

Hart et al. (2009) used the model of the University of Brighton's Community-University Partnership Program (CUPP) Helpdesk as an 'enabling platform' for commitment possibilities between the community and the university. The authors affirm that such models can help overcome the barriers holding back assistance to the community-university engagement work. Guodong and Huosong (2010) explain the role of knowledge as an important factor for the sustainable development of the university. Since the knowledge value is proven to slightly affect the knowledge dissemination, it is necessary to define the expectations behind the construction of local universities such as teacher quality, knowledge value, society effects, school conditions, the knowledge body and diffusion path.

Driouchi et al. (2008) describe the modern modes of information and knowledge diffusion to support regional decision-making processes in regions of Morocco. These modes can carry on innovation and local systems development that will assist in improving the competitiveness of a region. One of the mechanisms suggested is the establishment of local universities that would help in the transfer of knowledge from tacit to formal and assist in the dissemination of this latter. The suggested solution is derived from previous contributions of Driouchi and Djeflat (2004) and Driouchi et al. (2008). The new forms of regional universities result from the fact that local authorities and regions are not always involved in local educational systems despite their importance. Consequently, local populations are not involved enough in the knowledge production and diffusion processes. The regional dynamics are relevant to developing skilful and qualified human resources required to enhance local productivity and competitiveness. Therefore, the university has a fundamental role in local and regional development that might encompass the "second chance" education programs, which are open, flexible, and which create the conditions for excellence such as in France, the Netherlands, etc., polytechnic complexes such as in the US, India, China, etc., the "Land-Grant" programs in different parts of the US and the education schools managed by non-governmental institutions.

According to Driouchi and Zouag (2006), the local university in the regions of Morocco should be based on specific principles. These latter include the establishment of an open system, which recruits its students from different educational disciplines. No distinction exists between educational disciplines when recruiting students (i.e. technical, scientific, or literary streams, etc.) or among different types of educational systems (i.e. private, public, executive education, etc.). The local university provides a multi-disciplinary system including different domains of interest and education. It encourages the development of bridges and micro-bridges to allow students to change their specialty at different moments of the curriculum. The students receive training in different disciplines but can also pursue other studies according to their choices and interests. In the local university, education offerings and areas of specialty are adjusted to the trends expressed at the regional level. Another principle concerns the university complex that should educate students and their partners to develop a culture of tolerance, social work, and diversity of ideas and interests.

Figure 1: the Structure of a Regional University

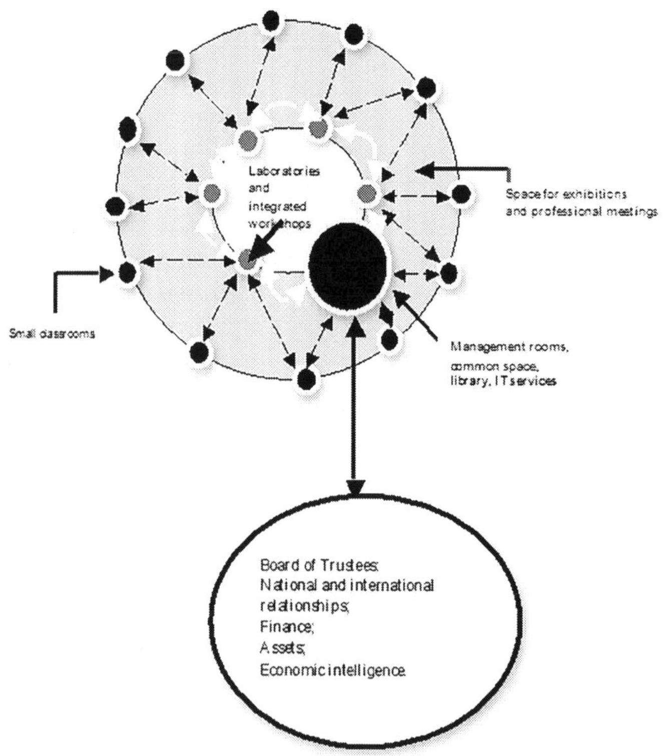

Source: Driouchi and Zouag, 2006.

Concerning the structure of these local universities, direct management of the university complex should involve three major poles, which are the financial, the admissions and the academic poles. The central entities of this system should include laboratories, integrated workshops and simulation centres. These entities will form the nodes around which several small classrooms will be designed. The common space is used for exhibitions and professional meetings as described in Figure 1.

The financial resources of the local university include revenues from its own assets (tangible and intangible assets developed and transferred to sustain the education and research system), donations given by the public administration, the private sector, and associations, income generated by products and services developed at the university, consulting revenues, tuition fees paid by the students, companies, and other institutions and revenues from the executive and continuing education programs and specific actions for the benefit of other institutions. The local university financial resources also include supplies from regional government or the state, revenues equivalent to fiscal exonerations, revenues related to general costs engaged as part of research and development projects and international donations.

Belenzon and Schankerman (2007) studied the effect of private ownership, motivation pay and local development objectives on university licensing performance. They found that using payment incentives impacts university licensing performance especially in the case of privately owned universities and schools. This is explained by the additional 30% to 40% income brought by license. They also found that around 30% less income by license is generated from universities that have strong local development objectives.

3. Empirical Analysis

This section introduces the discussion about the methods and data used before the introduction of the results attained.

Methods and Data

To test for the relationship between knowledge and local development measures, panels of regionalized data per country are needed.

In the absence of such data, two major implicit assumptions are set prior to the implementation of the empirical analysis using available information. The first assumption relates to considering that more universities in a given country would mean more regions covered by these sources of knowledge. The second assumption is related to the meaning of the overall development measures taken as the Human Development Index (HDI) and the Knowledge Economy Index (KEI) (WBI, 2011). These dependent variables are considered to be capturing larger regional and localized performances.

In practice and for the empirical assessment of the role of universities on local development, three groups of countries are considered to exhibit different patterns. These are developed countries with an important contribution of advanced knowledge, emerging economies with new roles of advanced knowledge but tacit knowledge is not absent, and developing countries where traditional knowledge is the most dominant. The distinction between countries in terms of development is based on their income level. Each of the three sets of countries is listed under developed, developing and emerging countries (see Tables in Appendices).

The data used to test the above hypotheses include the number of universities per country, the number of regions per country, the total population of a specific country and the total population of age 18 years old. The data for universities is retrieved from the Ranking Web of World Universities (2011).[1] The data for regions is obtained from Wikipedia.[2] The population data is from the World Bank (2010) and exceptionally from Wikipedia for some countries. To measure the local development, related variables are included such as the HDI, the KEI and other development indices. Other variables can also be included such as starts of new businesses, employment growth in urban areas and average earning across all jobs in the urban area. For the purposes of this study, the HDI is retrieved from the United Nations Development Program global report of 2010 (UNDP, 2010). Besides, the KEI is retrieved from the World Bank website from the knowledge for Development section. The data is divided into three sections that are the developed, developing and emerging countries.

Results of the empirical Analysis

After listing the countries included in each of the three sets, regression analyses are performed. The dependent variable is the HDI and the independent variable is the number of universities per region. Another independent variable can be the number of universities per population of age 18. The results of these analyses are summarized in Tables 1 and 2.

Table 1: Regression analyses results

Countries	Equation	R2	Obs.
Developed	$\ln(HDI2010) = \underset{(-13.226)}{-0.2407} + \underset{(4.0121)}{0.0174^*} \ln(NumOfUniversities)$	0.2678	46
Developing	$\ln(HDI2010) = \underset{(-2.134)}{-0.2917} - \underset{(-0.6665)}{0.0166^*} \ln(NumOfUniversities)$	0.1462	93
Emerging	$\ln(HDI2010) = \underset{(-2.134)}{-0.2917} - \underset{(-0.6665)}{0.0166^*} \ln(NumOfUniversities)$	0.0255	19

The results of the regressions show that the human development of economies is slightly explained by the number of universities in the given countries where all the data is in a logarithmic form. The portion explained by the number of universities demonstrates that an increase in that number for developed countries leads to the slow increase of the human development factor by approximately 0.02%. At the same time, an increase in the number of universities leads to a slow increase in the human development effect on developing countries. Emerging economies have statistically insignificant results related to the effect of the number of universities on the human development effect of the country.

Table 2: Regression analyses results (continued)

Countries	Equation	R2	Obs.
Developed	$\ln(HDI2010) = \underset{(-14.74)}{-0.161} + \underset{(1.803)}{0.023^*} \ln(UnivPerPop18)$	0.069	46
Developing	$\ln(HDI2010) = \underset{(-2.796)}{-0.124} + \underset{(13.49)}{0.203^*} \ln(UnivPerPop18)$	0.667	93
Emerging	$\ln(HDI2010) = \underset{(-4.769)}{-0.203} + \underset{(4.705)}{0.1^*} \ln(UnivPerPop18)$	0.566	19

From Table 2, another factor can explain the level of human development. It is the number of universities per population aged 18 that can significantly affects the 2010 human development index in developing and emerging markets. This means that an increase in the number of universities per population of 18 years old leads to slightly increasing rise (0.203%) in HDI while it contributes to a 0.1 % increase of HDI for emerging economies. The relationship for developed countries is found to be statistically insignificant.

Table 3 provides results for using the KEI as an independent variable and the number of universities as well as the number of universities per population aged 18 as dependent variables.

Table 3: Regression analyses with KEI data

Countries	Equation	R2	Obs.
Developed	$\ln(HDI2010) = \underset{(-13.226)}{-0.2407} + \underset{(4.0121)}{0.0174}^{*} \ln(NumOfUniversities)$	0.2678	46
Developing	$HDI2010) = \underset{(-14.335)}{-0.858} + \underset{(3.9468)}{0.082}^{*} \ln(NumOfUniversit$	0.1462	93
Emerging	$\ln(HDI2010) = \underset{(-2.134)}{-0.2917} - \underset{(-0.6665)}{0.0166}^{*} \ln(NumOfUniversities)$	0.0255	19

The results listed in Table 3 show that for developed countries, both the number of universities and the number of universities per population aged 18, significantly impact the level of KEI in a positive way. In their logarithmic form, an increase in the number of universities by 1% can lead to the increase of KEI level by 0.033% and an increase in the number of universities per population aged 18 by 1% can lead to the increase in the KEI level by 0.085%. For developing countries, the increase in the number of universities does not significantly impact the KEI level. However, the number of universities per population aged 18 does positively influence the level of KEI. An increase in the number of universities per population of age 18 leads to the increase of the KEI level by 0.362%. In the case of emerging economies, the analyses show that an increase in the number of universities per population of age 18 of 1% can lead to the increase in the KEI level by 0.186%. However, the increase in the number of universities does not have a significant impact on the KEI level (Table 3).

Discussion

Following the empirical analysis above, the results show that the development in the number of universities in the developed countries depicts a thorough local growth driven by the increase in the expansion of tertiary education. This growth shows the level of country development in terms of knowledge economy and human development. It is proven that the knowledge economy index of developed economies is high as well as their level of human development index. Concerning developing countries, two trends are observed. The first concerns the traditional ways that affect local development while the second witnesses the emergence of new sources of local growth such as the establishment of additional universities. The second approach shows that innovation take off can take a new form that it not distinguished by privatization of public stakes or only by the creation of enterprises. In the opposite the establishment of regional and local universities is handled locally by public authorities. This trend is more seen in developed and emerging economies than in developing ones.

In developing countries, development indices show declining performances and the move from old to new sources of local growth is not straightforward. Economies in transition, such as Brazil, India, China, South Africa and Turkey, have improved development indices (KEI and HDI) and their levels of response to knowledge measure is lower than those of developing economies but higher than those of developed countries. This might confirm that shift in regimes from old to new sources of local development is likely to be the major driver of better local and global development. The analyzed data in this paper did not especially focus on the way North African countries dealt with their innovation take off. However, it is safe to say, based on developing countries' analyses, for innovation to take place, it is important that these developing nations first provide the necessary conditions, which include decent education, reliable infrastructures and especially a basis for research and development. This latter is better provided through universities and research centres, based inside these localized universities.

5. Conclusion

The needs of different population groups may lead to major reforms in many sectors, such as education, health, finances and employment. To satisfy the majority of needs, local growth is a key factor that results specifically from the establishment of local universities in different countries, thus involving both rural and urban circles in the development process. This paper has shown the role played by universities in offering knowledge in order to generalize local growth and provide population needs. The regression analyses pursued have concluded that the establishment of universities is the reason behind high development indices of developed countries. These regressions have also shown that developing countries have declining development indices because of the low number of universities, while emerging countries have exhibited some improvement relative to the case of developing countries.

Furthermore, the attained results have indicated that developing countries have more rooms for new regional universities because of the higher sensitivity of their development indices to knowledge measures. In addition, these developing countries can use their local and regional universities, along with research and development centres to innovate more effectively. The dependence of local development on local university establishment and impact is based on the identification of each university region-specific niches and economic opportunities that would allow innovation to take off. This latter idea can be studied and demonstrated in future studies. It is also necessary for local governments to establish the prerequisites for such innovation take off through supporting education, employment conditions and compensation systems. The above analysis has been conclusive about the role of tertiary local education as a sure engine for local development. But, further data are needed to refine the outcomes of this paper.

Notes

1 See: http://www.webometrics.info/university_by_country_select.asp

2 See: http://en.wikipedia.org/wiki/Category:Regions_by_country

References

Amsden, A. (1989), *Asia Next Giant*, Ithaca, NY: Cornell University Press.

Anderson, B.A., Romani, J.H., Wentzel, M. and Phillips, H.E. (2010), 'Awareness of water pollution as a problem and the decision to treat drinking water among rural African households with unclean drinking water: South Africa 2005,' *Population Studies Center Research Report 10-701*, Population Studies Center, University of Michigan, May.

Belenzon, S. and Mark, S. (2007), 'The Impact of Private Ownership, Incentives and Local Development Objectives on University Technology Transfer Performance,' *CEP Discussion Papers, dp0779*, Centre for Economic Performance, London School of Economics, UK.

Beneria, L. (2010), 'Keynote Address: Globalization, Women's Work and Care Needs: The Urgency of Reconciliation Policies,' in *North Carolina Law Review*, 88 N. C. L. Rev. 1501.

De Pee, S., Brinkman, H-J., Webb, P., Godfrey, S., Hill, I. D., Alderman, H., Semba, R. D., Piwoz, E. and Bloem, M. W. (2010), 'How to Ensure Nutrition Security in the Global Economic Crisis to Protect and Enhance Development of Young Children and Our Common Future,' in *Journal of Nutrition*, January, 140(1):138S-42S.

Driouchi, A. and Djeflat, A. (2004), *Economie de la Connaissance au Maroc: Enjeux et Perspectives*, AUI Publications.

Driouchi, A., Baijou, A., Moustaghfir, K. and Zouag, N. (2008), 'A Preliminary Study for the Design and Implementation of innovative and Distributed Decision Making Models for local government in Morocco,' ISUFI- LECCE, Italy/ IEAPS- AUI, Morocco.

Driouchi, A. and Zouag, N. (2006), 'Regional Universities', *Working Paper*, The Institute of Economic Analysis and Prospective Studies (IEAPS), Al Akhawayn University, Ifrane, Morocco.

Freeman, C. (2004), 'Technological infrastructures and international competitiveness', in *Industrial and Corporate Change*, 13(3): 541-569.

Frenette, M. (2009), 'Do universities benefit local youth? Evidence from the creation of new universities,' in *Economics of Education Review*, 28(3): 318-328.

Guodong, C. and Xia, H. (2010), 'Models and positive research on local university teachers' knowledge diffusion,' in *Science Research Management*, 2010-05, Department of Computer and College of Economics and Management, China.

Hart, A., Northmore, Chloe Gerhardt, C. and Rodriguez, P. (2009), 'Developing Access between Universities and Local Community Groups: A University Helpdesk in Action,' in *Journal of Higher Education Outreach and Engagement*, 13(3): 45-60.

Hill, K. (2006), 'University Research and Local Economic Development,' Department of Economics and Center for Competitiveness and Prosperity Research, Arizona State University. Available at: http://www.asu.edu/president/p3/Reports/univ Research.pdfUH

Ji, Q., Dai, G. and Rong, W. (2010), 'The role of local colleges in scientific and technological innovation in China,' School of Education, Jiangsu Teachers University of Technology, Changzhou, China; Optic Photonics and Energy Engineering (OPEE), 10-11 May, 2010, pp: 429 – 33.

Kebede, E., Kagochi, J. and Jolly, C.M. (2010), 'Energy consumption and economic development in Sub-Sahara Africa,' in *Energy Economics*, 32(3): 532-537.

Lundvall, B-A. (2007), 'Higher Education, Innovation and Economic Development,' *Paper presented at The World Bank's Regional Bank Conference on Development Economics,* Beijing, China.

Mazzoleni, R and Nelson, R.R. (2005), 'The Role of Research at Universities and Public Labs in Economic Catch-up,' *Initiative for Policy Dialogue - Working Paper*, Columbia University, New York.

McTavish, S., Moore, S., Harper, S. and Lynch, J. (2010), 'National female literacy, individual socio-economic status, and maternal health care use in sub-Saharan Africa,' in *Social Science and Medicine*, 71(11): 1958-63.

Pavitt, K. (1984), 'Sectoral Patterns of Technical Change: Towards a Taxonomy and a Theory,' in *Research Policy*, 13(6): 343–373.

Peres, W. (2006), 'The (slow) return of industrial policies in Latin America and the Caribbean', *Initiative for Policy Dialogue - Working Paper*, Columbia University, New York.

Quisumbing, A., R. and Pandolfelli, L. (2010), 'Promising Approaches to Address the Needs of Poor Female Farmers: Resources, Constraints, and Interventions,' in *World Development*, 38(4): 581-592.

Reinert, E.S. (2007), *How Rich Countries Got Rich ... and Why Poor Countries Stay Poor*, New York: Carroll & Graf Publishers.

Reinert, S. E. (2006), 'Institutionalism Ancient, Old and New: a Historical Perspective on Institutions and Uneven Development,' *WIDER Research Paper 2006/77,* UNU World Institute for Development Economics Research (UNU-WIDER), Helsinki, Finland.

Rothwell, R. (1977), 'The characteristics of successful innovators and technically progressive Firms,' in *R&D Management*, 7(3): 191-206.

Shaffer, D., F., and Wright, D., J. (2010), 'A New Paradigm for Economic Development – How Higher Education Institutions Are Working to Revitalize Their Regional and State Economies,' The Nelson A. Rockefeller Institute for Government, The Public Policy Research Arm of the State University of New York, Albany, New York.

Stiglitz, J. (2001), 'More Instruments and Broader Goals Moving toward the Post-Washington Consensus,' in Chang, H. (ed.), *The Rebel Within*, London: Wimbledon Publishing Company, London.

UNDP – United Nations Development Program (2010), *Human Development Report 2010*, Available at: http://hdr.undp.org/en/.

Westnes, P., Hatakenaka, S., Gjelsvik, M. and Lester, R.K. (2009), 'The Role of Universities in Strengthening Local Capabilities for Innovation – A Comparative Case Study,' in *Higher Education Policy,* 22: 483–503.

World Bank (2010), 'Total Population,' in World Development Indicators (WDI) & Global Development Finance (GDF), *World databank,* Available at: http://data bank.worldbank.org/ddp/home.do?Step=1&id=4.

WBI – World Bank Institute (2011), 'Knowledge Economy index (KEI),' *Knowledge for Development Program,* Available at: http://info.worldbank.org/etools/kam2/KAM_page5.asp.

Acknowledgement

The contribution of the project "Economics of Intellectual Property in Morocco and Neighboring Countries" supported by the Hassan II Academy of Sciences & Technologies, Rabat, Morocco, is highly acknowledged.

APPENDICES

Table A.1: Developed Countries

Developed	Num. Univ.	Regions	Pop18	Univ/Pop18 (000)	HDI	KEI
Australia	91	9	289624	0.31	0.937	8.97
Austria	77		100240	0.77	0.851	8.91
Bahrain	14		12280	1.14	0.801	6.04
Barbados	1		4208	0.24	0.788	7.16
Belgium	100	4	126862	0.79	0.867	8.80
Canada	204	11	450917	0.45	0.888	9.17
Croatia	26	6	55049	0.47	0.767	7.28
Cyprus	15		13366	1.12	0.810	7.50
Czech Republic	57	16	130932	0.44	0.841	7.97
Denmark	95	7	66978	1.42	0.866	9.52
Estonia	35		19175	1.83	0.812	8.42
Finland	51	22	66371	0.77	0.871	9.37
France	581	30	761763	0.76	0.872	8.40
Germany	411	13	935663	0.44	0.885	8.96
Greece	64	10	116095	0.55	0.855	7.39
Hong Kong	26		87084	0.30	0.862	8.32

Developed	Num. Univ.	Regions	Pop18	Univ/Pop18 (000)	HDI	KEI
Hungary	75	11	127100	0.59	0.805	8.00
Iceland	9		4529	1.99	0.869	8.95
Ireland	50		58262	0.86	0.895	9.05
Israel	33	3	110145	0.30	0.872	8.01
Italy	203	24	575897	0.35	0.854	7.79
Japan	716	9	1244995	0.58	0.884	8.42
Korea, Rep.	398		663394	0.60	0.877	7.82
Kuwait	10		40021	0.25	0.771	5.85
Latvia	63		33946	1.86	0.769	7.65
Luxembourg	4		5749	0.70	0.852	8.64
Malta	3		5608	0.53	0.815	7.58
Netherlands	160	3	201427	0.79	0.890	9.35
New Zealand	45	23	63601	0.71	0.907	8.92
Norway	67	1	62518	1.07	0.938	9.31
Poland	433	9	546668	0.79	0.795	7.41
Portugal	111	4	116911	0.95	0.795	7.61
Qatar	4		10185	0.39	0.803	6.73
Saudi Arabia	43		485787	0.09	0.752	5.31
Singapore	18	5	66174	0.27	0.846	8.44
Slovakia	33	9	78424	0.42	0.818	7.47
Slovenia	32		23185	1.38	0.828	8.15
Spain	236		453738	0.52	0.863	8.28
Sweden	50		128089	0.39	0.885	9.51
Switzerland	107	7	94647	1.13	0.874	9.01
Trinidad and Tobago	5		25029	0.20	0.736	5.59
United Arab Emirates	36		57378	0.63	0.815	6.73
United Kingdom	233	3	806680	0.29	0.849	9.10
United States	3274	28	4514594	0.73	0.902	9.02

Table A.2: Developing Countries

Developing	Num. Univ.	Regions	Pop18	Univ/Pop18 (000)	HDI	KEI
Albania	24		63603	0.38	0.719	3.96
Algeria	46		753091	0.06	0.677	3.22
Angola	8		379066	0.02	0.403	2.00
Argentina	106		681822	0.15	0.775	5.57
Armenia	19	1	61315	0.31	0.695	5.65
Azerbaijan	36		189365	0.19	0.713	3.83
Bangladesh	96		3373443	0.03	0.469	1.48
Belarus	48	7	146132	0.33	0.732	4.93
Benin	10		196395	0.05	0.435	2.05

Developing	Num. Univ.	Regions	Pop18	Univ/Pop18 (000)	HDI	KEI
Bolivia	46		202613	0.23	0.643	3.46
Bosnia and Herzegovina	37	3	51450	0.72	0.710	4.58
Botswana	3		44108	0.07	0.633	3.88
Bulgaria	60		93802	0.64	0.743	6.99
Burkina Faso	2	13	323167	0.01	0.305	1.71
Cambodia	34		362275	0.09	0.494	1.56
Cameroon	12	11	418946	0.03	0.460	1.71
Cape Verde	3		12623	0.24	0.534	3.35
Costa Rica	58		89194	0.65	0.725	6.03
Cote d'Ivoire	4	19	446357	0.01	0.397	1.65
Djibouti	1	3	19311	0.05	0.402	1.47
Dominican Republic	33		186288	0.18	0.663	3.85
Ecuador	70		261550	0.27	0.695	3.90
El Salvador	28		133145	0.21	0.659	4.06
Ethiopia	34	11	1834106	0.02	0.328	1.30
Fiji	4		16530	0.24	0.669	4.20
Georgia	17	11	74864	0.23	0.698	5.21
Ghana	28	12	521663	0.05	0.467	2.46
Guatemala	22		290970	0.08	0.560	2.89
Guinea	1	8	197600	0.01	0.340	1.07
Guyana	1	10	11621	0.09	0.611	4.57
Honduras	13		162495	0.08	0.604	3.21
Iran	533		1777261	0.30	0.702	3.75
Jamaica	10		55482	0.18	0.688	4.90
Jordan	37		122997	0.30	0.681	5.54
Kazakhstan	116		308677	0.38	0.714	5.05
Kenya	44		833810	0.05	0.470	2.77
Kyrgyzstan	11		119132	0.09	0.598	4.29
Lao PDR	3		146135	0.02	0.497	1.94
Lesotho	1		49363	0.02	0.427	2.05
Lithuania	48		52911	0.91	0.783	7.77
Macedonia	11		31928	0.34	0.701	5.58
Madagascar	10	24	419530	0.02	0.435	2.21
Malawi	5		301417	0.02	0.385	1.69
Mauritania	2	14	65167	0.03	0.433	2.36
Mauritius	3		20567	0.15	0.701	5.48
Moldova	20		73617	0.27	0.623	5.07
Mongolia	16		61943	0.26	0.622	4.72
Mozambique	7		457049	0.02	0.284	1.58
Myanmar	4		936540	0.00	0.451	1.34

Developing	Num. Univ.	Regions	Pop18	Univ/Pop18 (000)	HDI	KEI
Namibia	4	15	52934	0.08	0.606	4.28
Nepal	20		620385	0.03	0.428	1.74
Nicaragua	31		128238	0.24	0.565	2.81
Nigeria	85		3275446	0.03	0.423	1.84
Pakistan	270	1	3949225	0.07	0.490	2.34
Panama	23		60305	0.38	0.755	5.16
Paraguay	43		131340	0.33	0.640	4.00
Romania	111		292561	0.38	0.767	6.43
Rwanda	5		245946	0.02	0.385	1.14
Senegal	7	15	273491	0.03	0.411	2.57
Sierra Leone	2		118263	0.02	0.317	0.96
Sri Lanka	30		337308	0.09	0.658	4.17
Sudan	35	6	826528	0.04	0.379	1.78
Swaziland	1		30350	0.03	0.498	2.78
Syria	20		446389	0.04	0.589	3.09
Tajikistan	6		169459	0.04	0.580	3.22
Tanzania	32	27	872082	0.04	0.398	2.17
Tunisia	44		211737	0.21	0.683	4.42
Uganda	17		695336	0.02	0.422	2.36
Ukraine	330	2	664120	0.50	0.710	6.00
Uruguay	14		52397	0.27	0.765	6.49
Uzbekistan	36		654852	0.05	0.617	3.25
Venezuela	58	1	544602	0.11	0.696	4.18
Vietnam	86		1919088	0.04	0.572	3.51
Yemen	12		536434	0.02	0.439	2.20
Zambia	8		269983	0.03	0.395	2.12
Zimbabwe	8		344314	0.02	0.140	2.25

Table A.3: Emerging Countries

Emerging Countries	Num. Univ.	Regions	Pop18	Univ/Pop18 (000)	HDI	KEI
Brazil	1379	1	3356043	0.41	0.699	5.66
Chile	82	20	302144	0.27	0.783	7.09
China	1156	4	22349307	0.05	0.663	4.47
Colombia	282		887875	0.32	0.689	4.84
Czech Republic	57	16	130932	0.44	0.841	7.97
Egypt	52		1556129	0.03	0.62	4.08
Hungary	75	11	127100	0.59	0.805	8.00
India	1555	22	23624450	0.07	0.519	3.09
Indonesia	342		4242607	0.08	0.6	3.29
Malaysia	77		521077	0.15	0.744	6.07
Mexico	906	9	2024917	0.45	0.75	5.33
Morocco	95	16	650413	0.15	0.567	3.54

Peru	93	25	572907	0.16	0.723	4.79
Philippines	289	18	1845449	0.16	0.638	4.12
Poland	433	9	546668	0.79	0.795	7.41
Russia	671		2099258	0.32	0.719	5.55
South Africa	26		983380	0.03	0.597	5.38
Thailand	150	1	977756	0.15	0.654	5.52
Turkey	162	7	1357648	0.12	0.679	5.55

Table A.4: Developed Countries (Logarithm form)

Developed	**Num. Univ.**	**Univ/Pop18 (000)**	**HDI**	**KEI**
Australia	4.51086	-1.16	-0.07	2.19
Austria	4.34381	-0.26	-0.16	2.19
Bahrain	2.63906	0.13	-0.22	1.80
Barbados	0	-1.44	-0.24	1.97
Belgium	4.60517	-0.24	-0.14	2.17
Canada	5.31812	-0.79	-0.12	2.22
Croatia	3.2581	-0.75	-0.27	1.99
Cyprus	2.70805	0.12	-0.21	2.01
Czech Republic	4.04305	-0.83	-0.17	2.08
Denmark	4.55388	0.35	-0.14	2.25
Estonia	3.55535	0.60	-0.21	2.13
Finland	3.93183	-0.26	-0.14	2.24
France	6.36475	-0.27	-0.14	2.13
Germany	6.01859	-0.82	-0.12	2.19
Greece	4.15888	-0.60	-0.16	2.00
Hong Kong	3.2581	-1.21	-0.15	2.12
Hungary	4.31749	-0.53	-0.22	2.08
Iceland	2.19722	0.69	-0.14	2.19
Ireland	3.91202	-0.15	-0.11	2.20
Israel	3.49651	-1.21	-0.14	2.08
Italy	5.31321	-1.04	-0.16	2.05
Japan	6.57368	-0.55	-0.12	2.13
Korea, Rep.	5.98645	-0.51	-0.13	2.06
Kuwait	2.30259	-1.39	-0.26	1.77
Latvia	4.14313	0.62	-0.26	2.03
Luxembourg	1.38629	-0.36	-0.16	2.16
Malta	1.09861	-0.63	-0.20	2.03
Netherlands	5.07517	-0.23	-0.12	2.24
New Zealand	3.80666	-0.35	-0.10	2.19
Norway	4.20469	0.07	-0.06	2.23
Poland	6.07074	-0.23	-0.23	2.00

Developed	Num. Univ.	Univ/Pop18 (000)	HDI	KEI
Portugal	4.70953	-0.05	-0.23	2.03
Qatar	1.38629	-0.93	-0.22	1.91
Saudi Arabia	3.7612	-2.42	-0.29	1.67
Singapore	2.89037	-1.30	-0.17	2.13
Slovakia	3.49651	-0.87	-0.20	2.01
Slovenia	3.46574	0.32	-0.19	2.10
Spain	5.46383	-0.65	-0.15	2.11
Sweden	3.91202	-0.94	-0.12	2.25
Switzerland	4.67283	0.12	-0.13	2.20
Trinidad and Tobago	1.60944	-1.61	-0.31	1.72
United Arab Emirates	3.58352	-0.47	-0.20	1.91
United Kingdom	5.45104	-1.24	-0.16	2.21
United States	8.09377	-0.32	-0.10	2.20

Table A.5: Developing Countries (Logarithm form)

Developing	Num. Univ.	Univ/Pop18 (000)	HDI	KEI
Albania	3.17805	-0.97	-0.33	1.38
Algeria	3.82864	-2.80	-0.39	1.17
Angola	2.07944	-3.86	-0.91	0.69
Argentina	4.66344	-1.87	-0.25	1.72
Armenia	2.94444	-1.17	-0.36	1.73
Azerbaijan	3.58352	-1.66	-0.34	1.34
Bangladesh	4.56435	-3.56	-0.76	0.39
Belarus	3.8712	-1.11	-0.31	1.60
Benin	2.30259	-2.98	-0.83	0.72
Bolivia	3.82864	-1.48	-0.44	1.24
Bosnia and Herzegovina	3.61092	-0.33	-0.34	1.52
Botswana	1.09861	-2.69	-0.46	1.36
Bulgaria	4.09434	-0.45	-0.30	1.94
Burkina Faso	0.69315	-5.09	-1.19	0.54
Cambodia	3.52636	-2.37	-0.71	0.44
Cameroon	2.48491	-3.55	-0.78	0.54
Cape Verde	1.09861	-1.44	-0.63	1.21

Developing	Num. Univ.	Univ/Pop18 (000)	HDI	KEI
Costa Rica	4.06044	-0.43	-0.32	1.80
Cote d'Ivoire	1.38629	-4.71	-0.92	0.50
Djibouti	0	-2.96	-0.91	0.39
Dominican Republic	3.49651	-1.73	-0.41	1.35
Ecuador	4.2485	-1.32	-0.36	1.36
El Salvador	3.3322	-1.56	-0.42	1.40
Ethiopia	3.52636	-3.99	-1.11	0.26
Fiji	1.38629	-1.42	-0.40	1.44
Georgia	2.83321	-1.48	-0.36	1.65
Ghana	3.3322	-2.92	-0.76	0.90
Guatemala	3.09104	-2.58	-0.58	1.06
Guinea	0	-5.29	-1.08	0.07
Guyana	0	-2.45	-0.49	1.52
Honduras	2.56495	-2.53	-0.50	1.17
Iran	6.27852	-1.20	-0.35	1.32
Jamaica	2.30259	-1.71	-0.37	1.59
Jordan	3.61092	-1.20	-0.38	1.71
Kazakhstan	4.75359	-0.98	-0.34	1.62
Kenya	3.78419	-2.94	-0.76	1.02
Kyrgyzstan	2.3979	-2.38	-0.51	1.46
Lao PDR	1.09861	-3.89	-0.70	0.66
Lesotho	0	-3.90	-0.85	0.72
Lithuania	3.8712	-0.10	-0.24	2.05
Macedonia	2.3979	-1.07	-0.36	1.72
Madagascar	2.30259	-3.74	-0.83	0.79
Malawi	1.60944	-4.10	-0.95	0.52
Mauritania	0.69315	-3.48	-0.84	0.86
Mauritius	1.09861	-1.93	-0.36	1.70
Moldova	2.99573	-1.30	-0.47	1.62
Mongolia	2.77259	-1.35	-0.47	1.55
Mozambique	1.94591	-4.18	-1.26	0.46
Myanmar	1.38629	-5.46	-0.80	0.29
Namibia	1.38629	-2.58	-0.50	1.45
Nepal	2.99573	-3.43	-0.85	0.55
Nicaragua	3.43399	-1.42	-0.57	1.03
Nigeria	4.44265	-3.65	-0.86	0.61
Pakistan	5.59842	-2.68	-0.71	0.85
Panama	3.13549	-0.96	-0.28	1.64
Paraguay	3.7612	-1.12	-0.45	1.39
Romania	4.70953	-0.97	-0.27	1.86
Rwanda	1.60944	-3.90	-0.95	0.13
Senegal	1.94591	-3.67	-0.89	0.94
Sierra Leone	0.69315	-4.08	-1.15	-0.04

Developing	Num. Univ.	Univ/Pop18 (000)	HDI	KEI
Sri Lanka	3.4012	-2.42	-0.42	1.43
Sudan	3.55535	-3.16	-0.97	0.58
Swaziland	0	-3.41	-0.70	1.02
Syria	2.99573	-3.11	-0.53	1.13
Tajikistan	1.79176	-3.34	-0.54	1.17
Tanzania	3.46574	-3.31	-0.92	0.77
Tunisia	3.78419	-1.57	-0.38	1.49
Uganda	2.83321	-3.71	-0.86	0.86
Ukraine	5.79909	-0.70	-0.34	1.79
Uruguay	2.63906	-1.32	-0.27	1.87
Uzbekistan	3.58352	-2.90	-0.48	1.18
Venezuela	4.06044	-2.24	-0.36	1.43
Vietnam	4.45435	-3.11	-0.56	1.26
Yemen	2.48491	-3.80	-0.82	0.79
Zambia	2.07944	-3.52	-0.93	0.75
Zimbabwe	2.07944	-3.76	-1.97	0.81

Table A.6: Emerging Countries (Logarithm form)

Emerging Countries	Num. Univ.	Univ/Pop18 (000)	HDI	KEI
Brazil	7.22911	-0.89	-0.36	1.73
Chile	4.40672	-1.30	-0.24	1.96
China	7.05272	-2.96	-0.41	1.50
Colombia	5.64191	-1.15	-0.37	1.58
Czech Republic	4.04305	-0.83	-0.17	2.08
Egypt	3.95124	-3.40	-0.48	1.41
Hungary	4.31749	-0.53	-0.22	2.08
India	7.34923	-2.72	-0.66	1.13
Indonesia	5.83481	-2.52	-0.51	1.19
Malaysia	4.34381	-1.91	-0.30	1.80
Mexico	6.80904	-0.80	-0.29	1.67
Morocco	4.55388	-1.92	-0.57	1.26
Peru	4.5326	-1.82	-0.32	1.57
Philippines	5.66643	-1.85	-0.45	1.42
Poland	6.07074	-0.23	-0.23	2.00
Russia	6.50877	-1.14	-0.33	1.71
South Africa	3.2581	-3.63	-0.52	1.68
Thailand	5.01064	-1.87	-0.42	1.71
Turkey	5.0876	-2.13	-0.39	1.71

CHAPTER 9

Constructing Multidimensional Education and Health Welfare Indexes in Cameroon: Multiple Correspondence Analyses

Boniface Ngah Epo & Francis Menjo Baye

1. Introduction

Debate around the issue of development is perceived as economic welfare enhancing changes that lead to an increase in human well-being brought about by innovative usage of resources that combat poverty and cause growth. Consequently, understanding mechanisms that dent welfare like inadequate human capital (education and health) can be important in formulating policy designs that can augment household well-being in a developing country setting like Cameroon. Therefore, developmental typologies should anchor on effective poverty reduction strategies skewed in favour of the hardest hit by poverty. Furthermore, these typologies should not only be in terms of monetary poverty, but also in terms of "lackings" in key human capital endowments like education and health as attempted in this paper. Evidence is now observed that most countries in collaboration with international institutions- World Bank, International Monetary Fund (IMF), United Nations, African Union, etc- are now strategizing on how to ameliorate welfare along monetary and non monetary dimensions (Epo et al., 2010).

Considering the ever persisting manifestations of poverty through inter- generational transfer of poverty, most developing countries are now pursuing growth objectives coupled with addressing concerns for the poor as enshrined in their respective Poverty Reduction Strategy Papers (PRSPs). In line with the above mentioned remarks, the government of Cameroon as indicated in the 2003 Poverty Reduction Strategy Paper (PRSP) and the 2009 Growth and Employment Strategy Paper (GESP) has been keen on ameliorating the welfare of its citizenry. In this regard, a review of the GESP indicates among others that, education and health are key issues on which the government of Cameroon has decided to target. These include, adjusting the education and health maps (notably through an efficient breakdown of investment), developing technical and technological education branches, upgrading education and training programs and drawing investment programs (GESP, 2009). This is important because education and health affect both the current and future utility and production function of an individual (Becker, 1964; Grossman, 1972). These individuals are generally household heads. Therefore, they inevitably affect household welfare.

Foregrounding the above remark, despite efforts by developing countries like Cameroon to reduce monetary poverty (Epo, 2010), we still observe the persistency of poverty. This is likely due to an inadequacy in complementing policy efforts and designs, by also investing in key human capital characteristics like education and health. For instance, poverty in Cameroon was virtually flat between 2001 and 2007, revolving around 40%, after decreasing by 13% between 1996 and 2001 (NIS, 2008). This observation calls for the necessity to understand other dimensions of poverty that can act as complementary findings to better understand why overall poverty has remained virtually flat in Cameroon between 2001 and 2007. In this regard, Sen (1999) indicates the necessity of complementing monetary and non-monetary analysis of poverty for efficient policy enactment. Likewise, we argue that Sen's (1999) capabilities approach calls for the veracity of understanding other dimensions of poverty. This is because, it is necessary to define and understand factors or modalities that affect the capabilities of individuals in attaining functioning. These capabilities are key human capital characteristics like education and health.

Resolving human capital deficiencies is expected to be instrumental in augmenting the standard of living of the population. Inadequate human capital like education and health may explain causes of poverty in a developing country setting like Cameroon. It is apparent that an initial mal-distribution of education and health as well as associated endowments should make it much harder for the poor to participate in, and therefore gain from, the process of economic growth. This will further dent any efforts in reducing poverty levels in a developing country like Cameroon. Investment in education, health and human capital formation leads to an increase in the potential earnings of economic agents and thus an essential catalyst for the national fight against poverty. Education increases the skills and productivity of poor households, enhances the wage level as well as the welfare of the population. For health, acquisition of good health contributes directly and indirectly to household utility and production functions, and consequently household welfare. The contribution of education and health to poverty reduction is generally found to be positive and significant when measured in monetary terms (Epo et al., 2010).

This study set as main research question, how do the nature of the relationship between modalities and key capabilities like education and health affect household economic well-being in Cameroon? Specific questions are: (1) what are the advantages of the MCA over other methods associated to factorial analysis and (2) which policies can be drawn from our findings. The main objective of this study is to evaluate the nature of the relationship between modalities and key capabilities like education and health that affect household economic well-being in Cameroon. Specific objectives are to: (1) review the concept of Multiple Correspondence Analyses (MCA) in the area of multidimensional poverty measurement and (2) obtain findings on the base of this methodology. The rest of the paper is outlined as follows, section two presents a brief review of literature, section three the methodology and data used in this study, section four results and section five suggest policy recommendation and conclusions.

2. Brief Review of Literature

A review of the origins of the Multiple Correspondence Analysis (MCA) identify that this technique has been popular in the French (Benzécri, 1977 and 1982; Lebart et al., 1977) and English literature (Greenacre, 1984; Lebart et al., 1984). However, the origin dates back to several apparently independent beginnings. We identify: (i) The Method of Reciprocal Averages (Richardson and Kuder, 1933; Horst, 1935); (ii) The Analysis of Variance Approach (Guttman, 1959; Nishisato 1979; 1980); (iii) the Principal component Analysis (PCA) Approach (Burt, 1953; Cazes et al., 1977) and; (iv) The Generalized Canonical Analysis Approach (Saporta, 1980; Leclerc, 1980). A detailed history of the first two approaches can be found in Nishisato (1980). Benzécri (1982) outlines the history of the Correspondence analysis. A synthesis of the last approach can be found in Leclerc (1980).

Regarding studies that evaluate multidimensional poverty, recently the MCA approach has been popularized in the field of multidimensional poverty analysis by Asselin (2002; 2005). A review of some studies indicate that several studies applying the MCA approach have been carried out in Senegal, Togo and Tunisia (Ki et al., 2005; Ayadi et al., 2007 and Lawson et al., 2007). The main idea that sprouts from these studies is that in keeping with Sen (1999) going beyond the monetary approach to investigate multidimensional poverty along dimensions like education, health and habitat constitute value addition in understanding poverty in Africa. In Cameroon, some studies apply the multidimensional concepts in analyzing poverty; Foko et al., (2007); Ningayé and Ndjanyou (2007); Njong and Ningaye (2008); Araar (2009) and Manga and Epo (2010). The main findings from these studies is that evaluating poverty status along non-income dimensions reveal that (a) poverty along non-income dimension is higher and more acute than monetary poverty; (b) multidimensional poverty is severe for women than men. It is also more acute in rural areas than urban areas; (c) policies that go beyond income to non-income dimensions like education, health and housing should be reinforced. The generation of MCA results in this paper differs from past studies that appllied this approach in Cameroon because it uses the recently published ECAM III household survey results carried out in 2007 and showed the "dynamics" observed with regard to the different modalities.

3. Methodology

Technically, MCA is obtained by using the standard Correspondence Analysis on an indicator matrix. In this situation, the percentage of explained variance need to be corrected, and the Correspondence Analysis inter-point distances need to be adopted (Abdi and Valentin, 2007). Each nominal variable can comprise several levels, and each of these levels is coded as a binary variable. MCA can also accommodate quantitative variables by recoding them as "bins". The coding schema of MCA implies that each row has the same total, which for CA implies that each row has the same mass.

The use of the MCA technique (see Asselin, 2005) requires two processes. One, we defined a unique numerical indicator, C, as a composite of k-primary indicators, which is computable for each population unit, U_i. Two, specify the MCA process proper. To answer the first requirement, let a population E of N-population units E_i be considered. We can describe an indicator by a set of k-primary indicators or variables, I_k, we want to measure with k >1. These indicators can be heterogeneous in nature (qualitative indicators e.g. household income, number of rooms, or qualitative or categorical e.g. ordinal variables like level of education or non-ordinal variables like occupation, geography). Note that, (a) we assumed that these variables are qualitative or categorical ordinal; and (b) the variables have a meaningful ordinal structure.

Defining a composite welfare indicator, C, takes the value C_i (I_{ik}, k=1) for a given element of the population unit, E_i, we proceed by either the data reduction technique or the PCA technique. These are constructed as basic techniques. The composite indicator is necessarily a reductive variable, since it tries to summarize k-variables into just one variable, which may potentially translate a lot of information. The Principle Component Analysis (PCA) was generally used as the basic optimal data reduction process. Unfortunately, limitations associated to this process justify the use of the MCA. In what follows: (1) we review the data reduction process with the PCA approach; and (2) showed the limitations. The PCA process builds a sequence of uncorrelated (orthogonal) and normalized linear combination of input variables, exhausting the whole variability of the set of variances. This variance is named total variance" and defined as the trace of their covariance matrix, thus the sum of the k variances. The basic idea is that we maximize the variance of scaled indicators (Burt, 1950; Greenace, 1984). The optimality in the PCA approach consists in that the first component has the maximal variance. The remaining components have decreasing variances. The sum of all these variances sums up to the total variance of the k indicators. Geometrically, the above process corresponds to a change in the Cartesian axis system of the k-dimensional Euclidian space R^k. This is computationally identical to considering the eigen values and eigen vectors. The eigen values relative to the factorial axis, determined by the eigen vectors gives us the variance. The first factorial axis is generally believed to constitute a good candidate as the considered welfare indicator if it fulfils a set of conditions (Asselin, 2005; Asselin and Anh, 2005).

Relying on the above explanation, for the population unit E_i, C can be expressed as:

$$C_i = \sum_{k=1}^{K} W^{i,k} I_i^{*k} \quad \text{.......... (1)}$$

where I_i^{*k} are the standardized primary indicators, $W^{i,k}$ the factor scores[1] or coefficients. They can be considered as multiple regression coefficients between the components and the standardized primary indicator. Considering the first component, as in this thesis, C_i is considered as the best regressed latent variable on the k-primary indicators.

The PCA approach has some limitations (Asselin and Anh, 2005; Moser and Felton, 2006). We note: (1) this technique was initially developed for a set of qualitative variables measured in the same unit. Unfortunately, the optimal sampling property for parameter estimates depend on the multivariate normal distribution and ceases to exist with the presence of qualitative variables. This limits the possibility of constructing representative composite variables; (2) the operationalization of the composite indicator for population units not involved in the sample used for estimation is not very appealing since weights are applicable to standardized primary indicators. These limitations naturally propel the second requirement, the MCA approach proper, which is judged as the most suitable factorial technique in constructing composite indexes and fulfil key properties to be outlined.

The Specificity of the Multiple Correspondence Analyses

To understand the value-added for choosing the MCA, let's assume that there exist K-primary indicators (Ik) with J categories that are ordinal (Jk). The total number of categories is $J = \sum_{k=1}^{K} J_k$. The various categorical ordinal variables can have several quantifications: (a) a binary indicator matrix; (b) a scaling of the categories; and (c) a scaling of the individuals. Associating to each primary indicator Ik the set of Jk binary variable 0/1, corresponding each to a category of the indicator, let's introduce the following notions, $X(N,J)$ the matrix of N observation on the K indicators decomposed into Jk variables. X is an indicator matrix. N_j : The absolute frequency of category j. This is the sum of the column j of the indictor matrix. N' : The sum of elements of the indicator matrix. $f_j = \frac{N_j}{N'}$: The relative frequency of category j. $f_j^i = {X(i,j)}/{X(i)}$: The conditional frequency of the indicator j, knowing the population unit, i. $X(i)$ is the sum of line, i, in the indicator matrix. The set $f_j^i = \{f_j^i, j = 1, J\}$ is named the profile of observation i.

We can now define the MCA as a PCA process applied to the indicator matrix, but with the c^2-metric on row/column profiles, instead of the usual Euclidean metric. The c^2-metric is a special case of the Mahalanobis metric. This metric has been used in the Generalized Canonical Analysis approach (Masson, 1974; Saporta, 1980). In the case of the MCA for the distance between two observed profiles *i and i'* in the R^j dimensional space takes the form:

$$d^2\left(f_J^i, f_J^{i'}\right) = \sum_{j=1}^{J}\left(\frac{1}{f_j} \right)\left(f_J^i - f_J^{i'}\right)^2 \qquad (2)$$

The difference between the Euclidean and indicator matrix resides in the term $\left(\frac{1}{f_j}\right)$ This term permits small categories to receive a higher weight in the computation of distances. When it comes to development analysis, the difference between the MCA and PCA regarding poverty (well-being) analysis is very meaningful concerning numerical results.

Asselin (2005) identified the properties of marginalization bias and reciprocal bi-addivity or duality, as key in showing the superiority to the MCA over the PCA in studying poverty issues. Regarding the first property, the MCA overweight the smaller categories within each primary indicator. For instance, in the case of a binary indicator, since smaller categories have same covariance as larger categories, their relative weights are higher. Concerning poverty analysis, attributing factorial weights to smaller groups, and arguing that we view poverty analysis in relative terms, these weights translate societal prioritization in resolving these issues. These weights can be expressed as:

$$W_{j_k}^{\alpha,k} = \frac{N}{N_{j_k}^{k}} \operatorname{cov}\left(F_\alpha^*, I_{j_k}^k\right) \qquad (3)$$

where $W_{j_k}^{\alpha,k}$ is the score of category jk on the non-normalized factorial axis α. $I_{j_k}^k$ the binary variable which takes the value 1 when the population unit has the category jk. F_α^* is the normalized score on the factorial axis α. $N_{j_k}^k$ is the frequency of the category jk of indicator k. cov is simply the covariance.

Concerning the second property, the concept of duality indicates that the MCA can be applied on the indicator matrix either to observations (row-profiles) or categories (column-profiles). This property can be expressed as:

$$F_{\alpha}^{i} = \frac{\sum_{k=1}^{K}\sum_{j_k=1}^{J_k}\frac{W_{j_k}^{\alpha,k}}{\lambda_{\alpha}} I_{i,j_k}^{k}}{K} \quad \text{.........} (4)$$

Where, K is the number of categorical indicators. Jk is the number of categories for indicator k. $W_{j_k}^{\alpha,k}$, the score of category Jk on the non-normalized factorial axis α. I_{i,j_k}^{k} the binary variable taking the value 1 when the unit i has the category Jk. F_{α}^{i} the score (non-normalized) of observation i on the factorial axis α. The above expression is for the row-profile of observation i. Reciprocally, the score of a category (column-profile) can be expressed as:

$$W_{j_k}^{\alpha,k} = \frac{\sum_{i=1}^{N_{Jk}}\frac{F_{\alpha}^{i}}{\lambda_{\alpha}}}{N_{j_k}^{k}} \quad \text{.........} (5)$$

More explicitly, the duality relation can be explained as follows. Assuming, for instance, that the first factorial axis meets the consistency conditions (see Asselin, 2005), we can consider this axis as the composite indicator of poverty or welfare- $C_i = F_1^i$. The basic idea here is that this axis most informs the individual on the behaviour of the different modalities (indicators) used to undertake this analysis. This relationship can be split into: (a-eq. 4) the composite poverty score of a population unit is the simple average of the factorial weights (standardized) of the K categories (poverty) to which it belongs. (b-eq. 5) the weight of a given poverty category is the simple average of the composite poverty scores (standardized) of the population units belonging to the corresponding poverty group.

Description of data used for this study and modalities used to construct the composite indexes for education and health

Data used in this study

The data used in this study is the ECAM II (2001) and ECAM III

(2007) Cameroon household consumption surveys. The ECAM II survey was undertaken from September to December 2001 (National Institute of Statistics, 2003). This household survey was carried out to remedy mistakes made in the first household survey and ameliorate information concerning the poverty profiles. This survey was comprised of about 12,000 households, of whom 10,992 were actually visited. In additions, (a) it was conducted to propose an adequate methodology for calculating poverty lines and profiles acceptable to major development partners, and which serves as a reference for further analysis; (b) To analyse monetary poverty, poverty in terms of living conditions of most households and potential poverty, while establishing the correlation between them; (c) Consolidate past analysis at a national and regional level, while isolating the two large towns (Douala and Yaoundé) and also distinguishing area of residence (urban or rural); and (d) To produce an adequate data base to ameliorate different statistics (of the population), notably in establishing household consumption in national accounts and updating data used in calculating price indexes (National Institute of Statistics, 2002a; 2002b).

The ECAM III survey was effectuated between May and July 2007; and comprised 11391 households. Its aim was to upgrade knowledge on poverty and welfare status in Cameroon by providing indicators that capture the living standards of the local population as to peg poverty profiles and act as a follow up of efforts made towards the implementation of the PRSP and the realization of the MDG objectives. According to the National Institute of Statistics, these data can be used to: (i) study all aspects of poverty at national and regional levels (monetary poverty, household poverty, poverty in terms of potentials and subjective poverty), as well as establish correlations between these different types of poverty; (ii) study the dynamics of poverty between 2001 and 2007, with the aim of evaluating the effects of macro-economic policies of the last five years on household wellbeing; (iii) evaluate the demand for education and identify its principal determinants; (iv) evaluate internal tourism in Cameroon and; (v) collect data of child labour in Cameroon (National Institute of Statistics, 2008).

After having justified the choice of the MCA, we will now look at the different modalities used to construct the composite index for education and health (as in Araar, 2009), and discuss some results. Concerning the synthetic variable-education, the following modalities (indicators) were chosen; knowing how to read and write, the household head has already attended school, first reason for dissatisfaction regarding the closest public primary school, first reason for dissatisfaction regarding the closest private primary school, distance to go to the nearest public primary school, distance to go to the nearest private primary school, required time to go the nearest primary public school, and required time to go the nearest private public school.

Some weighted descriptive statistics reveal that in 2001, 64% of households could read and write. For 2007, this value is 63%. In 2001 68% of household heads have at least gone to school. This value is two percentage points less than that for 2007. In like manner, 67% of households are situated less than one kilometre from public primary schools. For private schools, this percentage is 49%. The values for 2007 are two percentage points higher, 59% for 2001 and 68% for 2007 of households take at most 15 minutes to get to the nearest public primary school. In 2001 and 2007, 47% and 49% of these households take less than 15 minutes to get to the nearest private primary school, respectively. The normalized weighted average for the variable education is one for both years.

Regarding the synthetic variable-health, we chose the following modalities; sector of consultation, type of sanitary centre, appreciation of health status, first reason for dissatisfaction regarding the closest health centre, distance to go to the nearest sanitary centre, and required time to go the nearest health centre. Descriptive statistics for modalities used to construct the synthetic variable-health identify that 59% of household admitted that they usually consulted traditional practitioners in 2007. In 2001, this value was around 7%. About 59 percent of household lived less than three kilometres from the nearest health centre in 2007, whereas in 2001 this value was 63%. However, in rural areas the value is less than 16% in 2007 and 5% in 2001. In 2007, 63% of households take at most 25 minutes to get to the nearest health centre. In rural areas this value is about 32%. In 2001 this value was 58% for the general population and 3%

for rural areas. The normalized weighted average for the variable health is 1 for both years.

4. Results

The modalities used to construct the different synthetic variables respect the fundamental requirements outlined in Asselin and Anh (2005). They have an ordinal ordering consistent with their contributions in the first factorial axis (see Tables 1 and 2 in Appendices). The total number of observation, obtained from merging the ECAM I and ECAM II surveys, is 22 100 households. Results for the two synthetic variables identify that (see Tables 1 and 2 in Appendices): concerning the synthetic variable-education, made up of 28 dimensions, total inertia was 3.5. The different modalities reveal that, the ability to read and write contribute positively in increasing household economic well-being. In like manner, having schooled also behaves similarly. The modality distance to the nearest public and private school show that living less than 2kms helps household increase their well-being. This indicates that proximity to these facilities entails economies of scales with regards to time and money. These benefits are channelled towards other activities that can help household increase their well-being. For time, staying beyond 15 minutes from these facilities reduces household economic well-being.

Households identified that, concerning public schools, the two main reasons that causes difficulties in accessing these facilities are: (1) schools are generally expensive; and (2) these schools are far from their homes. This consequently decreases household well-being as captured by the synthetic variable-education. For private schools the main reason for dissatisfaction is that the schools are far. A key difference between these two modalities is that households do not consider private schools as being expensive. This is because they believe the relationship between the school fees paid in these schools equates the quality of education obtained.

Some main results associated to the synthetic variable-health (see Table 2 in Appendices) indicate that access to public health structures, private health structures and confessional health structures will increase household economic well-being. This composite variable has a total inertia of 4.8 and is made up of 29 dimensions. Perceiving health status as poor or average by household heads indicates that this is likely to reduce household economic well-being. Poor households admit that their health status is poor. This is probably because these households live in slumps where hygienic conditions are bad. Health constituting one of the important ingredients of human capital should be resolved to enhance collective welfare. Concerning the choice of the health structure, all categories of this modality contribute in increasing household economic well-being, except the category "going to the house of a medical doctor". The likely reason may be that for already impoverished households going to see a doctor for private medical consultation is very expensive, unlike public health structures where government grants some subventions. With regard to distance from the nearest health centre, living in less than four kilometres from these centres contribute in increasing household economic well-being. Distance and infrastructure leading to health centres are important. This is because poor households are likely to be situated far from health centres. Therefore, adding cost of transportation with health cost can only further adversely affect the welfare of poor households. As for time taken to the nearest district hospital, living in a house situated less than 15 minutes from the health centre contributes positively in household economic well-being.

5. Conclusion

There is sound intuitive, theoretical, and empirical reasons to believe that certain synergies exist between education, health, household economic well-being and poverty; partly rooted from the neo-classical human capital theory (Schultz (1961); Becker (1964)). Moreover, links between education, household economic well-being, and poverty is not infrequently cited in the policy literature and not least in Cameroon (Government of Cameroon, 2003).This paper uses the Multiple Correspondence analysis technique to investigate other dimensions of welfare, notably education and health, in Cameroon. The superiority of the multiple correspondence analyses over other methods is established. Results indicate that investing in educational infrastructure like quality of personnel, rending easy and rapid access to schools through reducing distance and time will inevitably increase household economic well-being and consequently reduce poverty. Likewise, economies of scales gained with regard to time and money are channeled towards other activities that can help household increase their well-being. Furthermore, we argue that though government policy for rendering free primary education for all is laudable, for this policy to have the expected effects, infrastructural policy design that targets issues the concord around efficient access to free education for all should also be considered in order to design efficient policy mix regarding education in Cameroon. Health constitutes a key component in welfare development. Results indicate that in addition to health infrastructure, other dimension of health like the quality of personnel, the motivation of these personnel and external infrastructure like roads and easy circulation also affect the health outcome. We also note that government's effort to render basic health facilities for all should be encouraged. However, hard and soft infrastructure that networks with this policy should also be targeted to enhance overall outcomes.

Concluding, and in line with Gary Becker, Grossman and Armatya Sen, human capital characteristics, represented in this study by the synthetic variables for education and health, constitute key elements in resolving deficiency in welfare, manifested by poverty and poor living standards. Furthermore, these characteristics are important regarding developing capabilities by individuals to help them break out of poverty traps.

Notes

1 They must have signs consistent with the interpretation of the 1st component, considered as a welfare indicator.

References

Abdi, H. and Valentin, D. (2007), 'Multiple Correspondence Analysis,' in Salkind, N. (ed.), *Encyclopaedia of Measurement and Statistics*, Thousand Oaks: Sage.

Araar, A. (2009), 'The Hybrid Multidimensional Index of Inequality', *Cahier de recherche/Working Paper 09-45*, Centre Interuniversitaire sur le Risque, les Politiques Economique et sur L'emploi (CIRPEE), October.

Asselin, L. M and Vu Tuan, A. (2005), 'Multidimensional Poverty in Vietnam 1993-1998, according to CBMS indicators', *Vietnam Socio-Economic Development Review,* No. 41, Spring.

Asselin, L.M. (2002), 'Multidimensional poverty Theory', *MIMAP Training Session on Multidimensional Poverty*, International Development Research Center, Quebec, Canada. June.

Asselin, L.M. (2005), 'Multidimensional Poverty Measurement with Multiple Correspondence Analysis', *Paper presented at the International conference: the Many Dimensions of Poverty*, Brasilia, Brasil, 29-31 August.

Ayadi, M., Lagha, E.L. and Chtioui, N. (2007), 'Pauvreté et Inégalité en Tunisie: une Approche Non-Monétaire', *Cahier de recherche, Poverty and Economic Policy Network-PMMA 2007-05.*

Becker, G. S. (1964), *Human Capital*, 1st edition, New York: Columbia University Press for the National Bureau of Economic Research.

Benzécri, J., P. (1977), 'Sur l'Analyse des Tableaux binaires Associés à une Correspondance Multiple', *Les Cahiers de L'Analyse des Données*, 2(1): 55-71.

Benzécri, J. P. (1982), 'Histoire et Préhistoire de l'Analyse des Données: l'Analyse des Correspondance', *Les Cahiers de L'Analyse des Données*, Paris : Dunod.

Burt, C. (1950), 'The Factorial Analysis of Qualitative Data', in *British Journal of Psychology*, 3: 166-185.

Burt, C. (1953), 'Scale Analysis and Factor Analysis,' in *British Journal of Statistical Psychology*, 6: 5-23.

Cazes, P., Baumerder, A., Bonnefous, S., and Pagés, J. P. (1977), 'Codage et Analyse des Tableaux Logiques. Introduction à la pratique des variables Qualitatives', *Cahiers du Bureau Universitaire de Recherche Opérationnelle* (Série Recherche, Cahier No. 27) Paris: Institut de Statistique des Universités de Paris, Universit6 Pierre et Marie Curie.

Epo, B. N. (2010), 'Moderate and Persistent Determinants of Poverty in Cameroon', Working paper series publications, Social

Science Research Network (SSRN), *Urban Economics and Regional Studies eJournal* (June).

Epo, B. N., Baye, F. M. and Manga, N.A.T. (2010), 'Explaining spatial and inter-temporal sources of poverty, inequality and gender disparities in Cameroon: a regression-based decomposition analysis', *Poverty and Economic Policy (PEP) Research Network,* Research project No. 11321 (June). Available at: www.pep-net.org.

Foko, B., Ndem, F. and Tchakote, R. (2007), 'Pauvreté et inégalités des conditions de vie au Cameroun: une approche micro multidimensionnelle', *Poverty and Economic Policy Research Network, Cahier de recherche PMMA*. Available at: www.pep-net.org.

Government of Cameroon, (2003), *The Poverty Reduction Strategy Paper*, Yaounde: Ministry of Economic Affairs, Programming and Regional Development.

Greenacre, M. J. (1984), *Theory and Applications of Correspondence Analysis*, London: Academic Press.

Grossman, M. (1972), 'On the concept of health capital and the demand for health', in *Journal of Political Economy*, 80(2): 223-255.

Government of Cameroon (2009), *Growth and Employment Strategy Paper*, Yaounde: Government Press.

Guttman, L. (1959), 'Metricizing rank-ordered or unordered data for a linear factor analysis', in *Sankhya*, 21: 257-268.

Horst, P. (1935), 'Measuring complex attitudes', in *Journal of Social Psychology*, 6: 369-374.

Ki, J. B., Faye, S. and Faya, B. (2005), 'Pauvreté Multidimensionnelle au Sénégal. une Approche Non-Monétaire par les besoins de Base', *Cahier de recherche, PEP-PMMA 2005-05.*

Lawson, B.B.K., Baniganti, K., Homevoh, E. and Lamadokou, E.A. (2007), 'Analyse Comparative de la Pauvreté et Inégalité au Togo: Une Approche Multidimensionnelle base su l'indice de Richesse', *Cahier de recherche, PEP-PMMA 2007-10.*

Lebart, L., Morineau, A. and Tabard, N. (1977), *Techniques de la Description Statistique*, Paris: Dunod.

Lebart, L., Morineau, A. and Warwick, K.M. (1984), *Multivariate Descriptive Analysis: Correspondence analysis and related Techniques for large Matrix*, New York: Wiley-Interscience.

Leclerc, A. (1980), 'Quelques Propriétés Optimales en Analyse de Données en terme de corrélation entre variables', in *Mathématique et Sciences Humaines*, 18: 51-67.

Manga, N.T.A and Epo, N.B. (2010), 'Pauvreté Multidimensionnelle au Cameroun: Une Alternative par L'Analyse en Composante Principales', in *African Review of Integration,* 4(1).

Masson, M. (1974), 'Processus Linéaires et Analyse des Données Non Linéaires', *Unpublished Doctoral Dissertation*, Université Pierre et Marie Curie, Paris.

Moser, C., and Felton, A. (2006), *The Construction of an Asset Index: Measuring Asset Accumulation in Ecuador*, Brooking Institution/ University of Maryland, Mimeo.

National Institute of Statistics, (2008), *Tendance, Profil et Déterminant de la Pauvreté au Cameroun en 2007.* Disponible sur *www. statistic-Cameroun.org*, May 2009.

National Institute of Statistics, (2002a), *CHCS II: Document de Méthodologie*, Yaoundé.

National Institute of Statistics. (2002b), *Evolution de la Pauvreté au Cameroun entre 1996 et 2001*, Decembre, 2002.

Ningayé, P. and Ndjanyou, L. (2007), *La Pauvreté Multidimensionnelle au Cameroun: sa Répartition Spatiale ses Déterminants*, Rapport Final, CREA, Kenya.

Nishisato, S. (1979), 'Dual Scaling and its variants', in *New Directions for Testing and Measurement*, 18(4): 1-12.

Nishisato, S. (1980), *Analysis of Categorical Data: Dual Scaling and its Applications*, Toronto: University of Toronto Press.

Njong, A.M and Ningaye, P. (2008), 'Characterizing weights in the measurement of multidimensional poverty: an application of data-driven approaches to Cameroonian data', *OPHI Working Paper 21*, Available at: www.ophi.org.uk.

Richardson, M., and Kuder, G. F. (1933), 'Making a Rating scale that Measures', in *Personnel Journal*, 12: 36-40.

Saporta, G. (1980), 'About some remarkable properties of generalized canonical analysis', *Paper presented at the Second European Meeting of the Psychometric Society*, Groningen, The Netherlands, June.

Schultz, T. W. (1961), 'Investment in Human Capital: Reply', in *The American Economic Review*, 51(5): 1035-1039.

Sen, A. K. (1999), *Development As Freedom*, Oxford: Oxford University Press.

APPENDICES

Table 1: Results for the Computation of the Synthetic Variable-Education

Number of Observations: 22001
Total Inertia: 3.5
Number of axis: 2

Variable	Scores		Correlation		Contributions	
	First Axis	Second Axis	First Axis	Second Axis	First Axis	Second Axis
Knowing how to read and write					0.123	
Yes	0.971	0.181	0.566	0.015	0.044	0.001
No	-1.746	-0.326	0.566	0.015	0.079	0.002
Household head has already attended school					0.119	
Yes	0.846	0.118	0.55	0.008	0.036	0.001
No	-1.95	-0.272	0.55	0.008	0.083	0.001
Distance to go to the nearest public primary school					0.04	

Less than 500 meter	0.273	-0.644	0.014	0.058	0.002	0.009
500<D<1km	0.44	0.014	0.034	0.000	0.005	0.000
1km<D<2km	0.013	0.605	0.000	0.017	0.000	0.003
2km<D<3km	-0.906	1.177	0.022	0.028	0.004	0.006
3km<D<4km	-1.923	0.72	0.03	0.003	0.006	0.001
4km<D<5km	-1.814	-0.494	0.02	0.001	0.004	0.000
D>6km	-2.588	0.951	0.094	0.01	0.019	0.002
Distance to go to the nearest primary public school					0.122	
Less than 500 meter	1.353	-0.164	0.284	0.003	0.042	0.001
500<D<1km	0.946	0.622	0.066	0.022	0.012	0.004
1km<D<2km	0.222	0.698	0.001	0.01	0.000	0.002
2km<D<3km	-0.511	1.142	0.004	0.017	0.001	0.004
3km<D<4km	-0.779	1.057	0.004	0.006	0.001	0.001
4km<D<5km	-1.306	-1.183	0.015	0.009	0.003	0.002
D>6km	-1.617	-0.469	0.441	0.029	0.063	0.005
Required time to the nearest primary public school					0.043	
0-5min	0.247	-0.946	0.006	0.066	0.001	0.013

6-15min	0.495	-0.202	0.06	0.008	0.008	0.001
16-25min	0.138	0.627	0.001	0.016	0.000	0.003
26-35min	-0.723	1.074	0.021	0.035	0.004	0.0008
36-45min	-1.281	1.147	0.024	0.015	0.005	0.004
> 46min	-2.236	0.733	0.122	0.01	0.025	0.002
Required time to the nearest primary private school					0.125	
0-5min	1.506	-0.381	0.18	0.009	0.031	0.002
6-15min	1.073	0.266	0.159	0.008	0.024	0.001
16-25min	0.04	0.625	0.000	0.011	0.000	0.002
26-35min	-0.486	0.84	0.006	0.014	0.001	0.003
36-45min	-1.343	0.503	0.026	0.003	0.005	0.001
> 46min	-1.718	-0.497	0.424	0.027	0.064	0.005
First reason for dissatisfaction regarding the closest public primary school					0.002	
Other reasons	0.115	-1.66	0.005	0.835	0.001	0.094
Very costly	-0.596	0.384	0.001	0.000	0.000	0.000
Too far	-0.126	1.991	0.004	0.831	0.001	0.113
First reason for dissatisfaction regarding the closest private primary school					0.003	
Other reasons	0.088	-1.691	0.003	0.825	0.000	0.096
Very costly	0.781	0.732	0.008	0.005	0.002	0.001
Too far	-0.172	1.995	0.008	0.787	0.001	0.11
TOTAL					**0.577**	

Source: Computed by authors using STATA 10

Table 2: Results for the Computation of the Synthetic varaible-Health

Number of Observations:22001
Total Inertia: 4.83
Number of axis:2

Variables	Scores		Correlation		Contributions	
	First Axis	Second Axis	First Axis	Second Axis	First Axis	Second Axis
Sector of consultation					0.229	
Public	1.216	0.693	0.317	0.074	0.054	0.015
Lay Private	0.494	-1.375	0.03	0.17	0.006	0.04
Missionary	1.187	-0.414	0.079	0.007	0.018	0.002
Others	-2.098	0.408	0.857	0.023	0.151	0.005
Type of sanitary centre					0.221	
1st class Hospitals	1.206	2.264	0.017	0.043	0.004	0.013
Provincial Hospitals	1.348	1.538	0.038	0.035	0.009	0.01
District Hospitals	1.218	0.135	0.093	0.001	0.021	0.000
Divisional Health Centre	1.248	-0.264	0.038	0.001	0.009	0.000
Integrated Health Centre	1.135	0.489	0.131	0.018	0.027	0.004
Pharmacy	0.833	0.227	0.004	0.000	0.001	0.000
Clinic	0.931	-0.568	0.021	0.006	0.005	0.002
Private Doctor	0.873	-1.641	0.007	0.017	0.002	0.005
Traditionalist	-1.924	0.277	0.839	0.013	0.14	0.002
Others	0.476	-2.394	0.012	0.220	0.003	0.059
Appreciation of health status					0.037	
Very Good	-0.555	0.091	0.115	0.002	0.016	0.000
Good	0.227	-0.179	0.008	0.003	0.001	0.001

Average	0.683	0.03	0.05	0.000	0.01	0.000
Bad	1.103	0.000	0.034	0.000	0.008	0.000
others	-2.348	-0.56	0.007	0.000	0.002	0.000
First reason for dissatisfaction regarding the closest sanitary centre					0.146	
Other reasons	-1.081	0.203	0.551	0.014	0.067	0.002
Very costly	0.083	0.761	0.000	0.006	0.000	0.002
Too far	1.292	-0.3	0.552	0.022	0.079	0.004
Distance to go to the nearest sanitary centre					0.011	
Less than 500 meter	0.308	1.997	0.011	0.328	0.002	0.078
500<D<1km	0.069	1.057	0.000	0.081	0.000	0.020
1km<D<2km	0.288	0.043	0.005	0.000	0.001	0.000
2km<D<3km	0.236	-0.447	0.002	0.006	0.001	0.002
3km<D<4km	0.158	-0.738	0.001	0.009	0.000	0.003
4km<D<5km	-0.638	-1.069	0.012	0.024	0.003	0.007
D>6km	-0.399	-1.866	0.024	0.375	0.004	0.084
Required time to go the nearest sanitary centre					0.006	
0-5min	0.214	2.078	0.004	0.246	0.001	0.063
6-15min	0.126	1.001	0.003	0.143	0.001	0.029
16-25min	-0.234	-0.184	0.004	0.002	0.001	0.000
26-35min	0.257	-0.924	0.003	0.03	0.001	0.008
36-45min	-0.079	-1.179	0.000	0.029	0.000	0.008
> 46min	-0.289	-2.047	0.01	0.361	0.002	0.085
TOTAL					**0.650**	

Source: Computed by authors using STATA 10

CHAPTER 10

An Investigation into the Factors Affecting Corporate Sustainability of Public Further Education and Training Colleges in Gauteng Province of South Africa

Manager M. Muswaba & Zeleke Worku

1. Introduction

Identification and implementation of factors affecting corporate sustainability serves to promote sustainable development in the public sector. The purpose of the paper is to investigate factors affecting corporate sustainability within the Further Education and Training Colleges (FET) Colleges sector. Sustainability is fundamental to the continuing operation of any organisation including FET Colleges and is arguably the fashionable concept of the moment. The paper serves to demonstrate that corporate sustainability is fundamental to the continuing operation of FET Colleges although less attention has been paid to the public sector. Both qualitative and quantitative approaches were used to collect data. The central part of this paper is based on exploration of the factors affecting sustainability by investigating the FET Colleges in Gauteng Province of

South Africa. The research analysis revealed that there were some factors where colleges had strengths and this became cause for optimism while there were factors which colleges did not even observe thereby being cause for concern. Areas where further work is needed were identified as well. The paper increased the understanding that factors that affect corporate sustainability and sustainable development in private organisations were the same as those that also affected sustainability of public institutions such as FET colleges.

The public sector and in particular Further Education and Training sector has received no attention in terms of empirical studies on sustainable development. However, there has been a growing support from the South African government to FET College sector evidenced by increased funding as well as providing governance autonomy to college council boards. While it is clear what is generally meant by good corporate governance it is much less clear what is meant by sustainable governance and as such the research aimed at investigating this concept with respect to Public FET Colleges in the Gauteng province.

The entire part of the paper would be in six parts. Following the introduction is a review of the existing literature pertinent to both the objectives and the research question. This is further followed by a description of the research design. The study then presents the results of the survey which in turn is followed by a discussion of the various sustainability factors. Finally, suggestions for further research concluded the discussion section.

2. Literature Review

Imagine the long-term leverage if higher education conducted the research for and helped to implement sustainability programs on campus and surrounding communities. Graduating students could then bring knowledge, skills and values of sustainability to their future employment, consumption decisions, lifestyle choices, and to the improvement of communities in which they live. As FET college corporate sustainability becomes more popular, it will no doubt become professionalised, leading to a discussion of the future path of sustainability on public FET colleges.

In 1987 the United Nations (UN) World Commission on Environment and Development issued the Brundtland Report, also known as 'Our Common Future', in response to the global environment crisis. The report defined sustainable development as that which implies meeting the needs of the present without compromising the ability of future generations to meet their own needs, (Report of the World Commission on Environment and Development, 1987). This definition of sustainability is the most widely accepted and is used here as a basis for understanding the modern concept of corporate sustainability.

To fill the gap between sustainable development and corporate sustainability, the Global Reporting Initiative (GRI) evolved in 1997 between partnerships with the Coalition for Environmentally Responsible Economies as a result of the Exxon Valdez 1989 oil spill. The main purpose of the GRI was to develop a standardised corporate reporting format based on the concept of sustainability, and to promote transparency of performance. Sustainability therefore is now an emerging model or paradigm of the current and future generations that addresses various stakeholder demands. Corporate sustainability attempts to align the self-interest of individuals and companies in a longer-term view using markets and eco-efficiency to drive innovation while addressing social and environmental concerns.

Elkington (2001) defined corporate sustainability, or triple-bottom line as a focus towards economic prosperity, environmental quality, and social justice. However, Welford and Jones, (1998) state that '... strategies are needed to translate conceptual theories of what sustainable development means into practical ways of achieving it over time within a corporate context...requiring them [companies] to look at their own ethics, their objectives and their own forms of organization, corporate culture and communication'.

The study of corporate sustainability is a young discipline that draws origins from the study of corporate social responsibility. Both fields view business responsibilities in a broader context than the narrower economic perspective of profit maximisation for shareholders. Therefore, a general claim to the corporate social responsibility development is a more inclusive form of management which Freeman (1984) referred to as stakeholder management. According to Freeman (1984) stakeholder management referred to any individual or group who affect or were affected by the actions, decisions, policies, practices or goals of an organisation.

It is the first time that stakeholder relationship management as part of the inclusive approach to corporate sustainability was treated as a separate chapter in King III. While initially it was considered part of sustainability in the previous report, it became clear during the deliberations on the various topics covered by King III that the management of relationships with stakeholders was such an important element of corporate governance and therefore needed more detailed coverage. It was concluded that pro-actively focusing on stakeholders with legitimate expectations and managing the relationships with them will contribute towards transparency and enable them to determine the value of public colleges and ultimately resulting in a positive reputation for these colleges.

David Saiia (2007) highlighted the importance of taking into account an organisation's stakeholders when making sustainability decisions. As Saiia noted, organisations that depend on community support can be influenced by the salience, or power, of key stakeholders. Public education institutions like FET colleges depend largely on government and community acceptance and support. These institutions therefore need to take into account their stakeholder. The fact that stakeholders interact with the colleges on multiple levels affecting people, planet and prosperity, further reinforces the need for stakeholder support and buy-in (Saiia, 2007).

Barlett and Chase (2004) stated that clarity about what constitutes sustainability for a particular institution will require the shared vision of many points of view. Financial pressures and the fact that every institution must cater to multiple stakeholders often slows or stalls colleges' sustainability movements. In order for a sustainability proposal to be put into action, stakeholders must be in agreement. This includes students, campuses, administrators, the board of council members and their major stakeholder, the state government. Because they provide the institution's funding, each stakeholder group must view sustainability as a top priority.

The multitude of sustainable practices that are being adopted all over the world makes up what author Edwards called the 'Sustainability Revolution,' The Sustainability revolution draws its significance and global impact from a wide spectrum of interests. According to Edwards (2005): 'Like its industrial counterpart, the Sustainability Revolution is creating a pervasive and permanent shift in consciousness and worldview affecting all facets of society ...The Sustainability Revolution draws its significance and global impact from a wide spectrum of interests with common fundamental values'.

Edwards categorised the core of modern sustainability into three parts namely ecology or environment, economy or employment, and equity or equality (Edwards, 2005). Edwards said that the three were made even more powerful by an active commitment to public education because education was a vehicle for helping everyone understands the dynamic nature of corporate sustainability. It is also through education that sustainability can become firmly established within the existing value structure of societies. Education, then, has the potential to be a central medium for change in any society.

A two-way symmetrical approach between an organisation and its stakeholders is also encouraged in King III. This is evident in chapter eight Mervyn (2009) on Managing Stakeholder Relationships, which states that: 'A stakeholder-inclusive corporate governance approach recognises that a company has many stakeholders that can affect the company in the achievement of its strategy and long-term sustained growth'. Crowther (2002) identified five aspects of sustainability which need to be recognised and analysed and these are:

- *Societal Influence*, which was defined as a measure of the impact that society makes upon the corporation in terms of the social contract and stakeholder influence;
- *Environmental Impact*, which was defined as the effect of the actions of the corporation upon its geophysical environment;
- *Organisational Culture*, which was defined as the relationship between the corporation and its internal stakeholders, particularly employees, and all aspects of that relationship and,
- *Finance*, which was defined in terms of an adequate return for the level of risk undertaken.

- *Stakeholders:* the community which has influence over the continuous and cause for survival of the organisation.

These five must be considered as the key dimensions of sustainability, all of which are equally important. They can also be resolved into a two-dimensional matrix along the polarities of internal versus external focus and short term versus long term focus, which together represent a complete representation of organisational performance. It is however fundamental to recognise the realities of the global environment in so far as the FET Colleges are firmly embedded into a global environment which necessarily takes into account the past and the future as well as the present. The figure below illustrates the five factors as given by (Crowther, 2002).

Figure 1: Mental Model of Corporate Sustainability

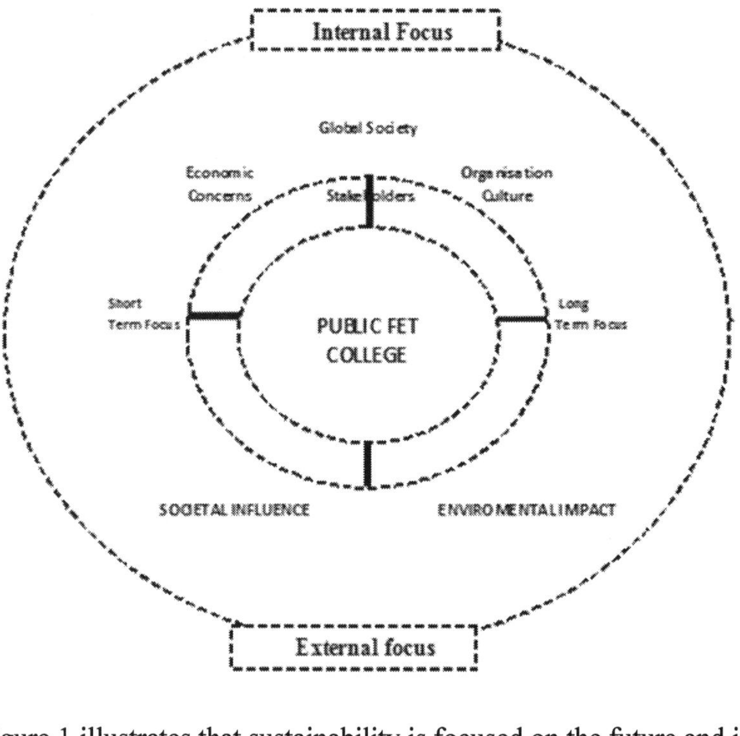

Figure 1 illustrates that sustainability is focused on the future and is concerned with ensuring that the choices of resource utilisation in the future are constrained by decisions taken in the present. This implies such concepts as generating and utilising renewable resources, minimising pollution and using new techniques of service and products distribution. Not only does such sustainable activity however impact upon society in the future but it also impacts upon the organisation and the environment it operates in the future. This means that good environmental performance by an FET college in the present is in reality an investment in the future of the college itself (Waddock and Graves, 2002). This is achieved through ensuring service provision techniques which will enable the colleges to operate in the future in a similar way to its operations in the present and also undertaking value creation activity in the future much as it does in the present.

College councils and college management should be concerned not only with arguments about resource distribution but rather with the development of measures which truly reflect the activities of the college upon its environment. These techniques of measurement, and consequently of reporting, are a necessary precursor to the concern with the management for the future, hence with sustainability. Similarly, the creation of value within the college should be followed by the distribution of value to the stakeholders of any college.

Value however must be viewed to include more than economic value as it is possible that economic value can be created at the expense of other constituent components of welfare such as spiritual or emotional welfare. This creation of value by the colleges should add to quality welfare for society at large, although this welfare could be targeted at particular members of society rather than treating all as equals. This has however led to arguments by Tinker (1988), Herremans *et al.* (1992) and Gray (1992), amongst others, concerning the distribution of value created and to whether value is created for one set of stakeholders at the expense of others. Nevertheless, if, when value is created it should add value to welfare for society at large, no matter how it is distributed as argued by Herremans *et al.* (1992).

Similarly, good environmental performance leads to increased welfare for society at large, and this tends to be expressed in emotional and community terms rather than being capable of being expressed in quantitative terms. This is expressed in a feeling of wellbeing, which will of course lead to increased motivation. Such increased motivation will inevitably lead to increased productivity, some of which will benefit the colleges themselves. The desire to maintain the pleasant environment will also eventually lead to a further enhanced environment, a further increase in welfare and the reduction of destructive aspects of societal engagement by individuals. Thus increased welfare would lead to its own self-perpetuation.

Werback (2009) added one more facet which is needed and this is the cultural aspect. Where the social aspect has to do with internal aspects such as employee working conditions and corporate contributions to social causes, the cultural bottom line has to do with the preservation and ongoing viability of distinct cultures, or ways of life, which may come under threat by such issues as rural urban migration. The model above is not static or stable as it might appear. Quite to the contrary, the model aims to depict sustainability as an ongoing process taking place between ever-changing, self-organising systems (Hawkin, 2007). Such complex systems and their interrelations are characterised by continual change and development, ongoing feedback across micro and macro levels, and by evolving bottom-up and top-down development.

Based on the model presented above, an approach to improving sustainability from within FET Colleges can be promoted. For the best chances of success, it is important to consider this process to be an action science approach to system-wide organisational learning. Argyris (1982) noted that complex conditions require that management shift from directing or dominating strategy to actively cultivating and enriching conditions for knowledge assimilation. Public institutions have always had weak corporate spirit. Therefore, this research aims to understand the development of corporate sustainability in the public FET colleges in South Africa. There is evidence of lack of research focusing on public colleges' sustainable development which should provide incentive for a full recognition of the systems, not only in FET colleges of Gauteng Province, but also in other provinces of South Africa and the world at large. Only when there is a clear understanding can college policymakers and the state make corporate sustainable more efficient and viable future.

3. Statement of the problem and research questions

FET colleges and all their stakeholders are responsible for sustainability. As this indicates the necessity for joint efforts to advance sustainable development, attempts are made to integrate the responsibilities for sustainability of organisations, society, consumers, and policy makers. It is also believed that views on corporate sustainability and corporate governance should be perceived as an alarm bell which board of directors need to consider in the process of stakeholder motivation in order to secure the college success in service delivery.

The meaning of organisational success and the sustainability principle suggests that the notion of indefinite organisational growth should be replaced with the notion of indefinite organisational development. This view on corporate sustainability advocates a broader scope of boards' decision-making strategies. Such governance tradition of sustainability considers simultaneous integration in organisational performance of the principles of economic welfare, social equity, and environmental preservation. If FET colleges operate in line with the sustainability principles they are assured of a balance of economic, social, and environmental sustainability.

The study investigated the following core research question as part of exploring the factors that affect corporate sustainability of FET Colleges: What are the factors that are essential for realising best practice and enforcing effective corporate sustainability of public FET Colleges in the Gauteng province of South Africa?

The specific sub-research questions for the study are given as:

- What factors lead to the successful implementation and ongoing success of corporate sustainability programs and policies in FET Colleges?
- Which models are suitable for ensuring sustainable development in the Further Education and Training Colleges in South Africa?
- What is the impact of corporate sustainability in the public FET institutions?

4. Methodology

Complete researches do always include important skills of gathering, processing and interpreting data through conducting and evaluating activities (McNabb, 2002). Adams further described it as a diligent search, studious inquiry, investigation or experimentation aimed at the discovery of new facts and findings (Adams *et al.,* 2007). This research considered important strategic methods scientifically which included defining the subject, collecting the most reliable data, ways of dealing with the data collected and analysed, and how to explain the result in a written form. For the purpose of this paper, the four steps approach of research design adopted is shown by Figure 2.

Figure 2: Four-steps Approach of the Research Methodology

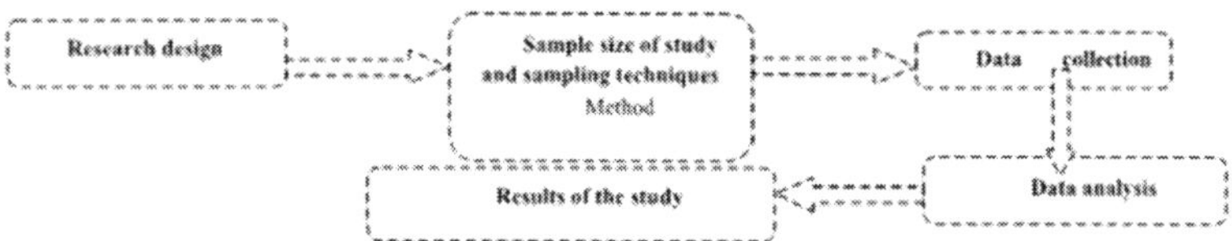

Research design

The design of the study was descriptive and cross-sectional. It was descriptive as the purpose of the study was to explore and describe factors that affect corporate sustainability. The design adopted the cross-sectional approach as data was gathered from respondents only once.

Sample size of study and sampling techniques

The population for the study comprised of all College Council members of public FET colleges in Gauteng province. Figure 3 illustrates the locations of FET colleges in Gauteng and the geographical spread of learners.

Figure 3: Locations of FET Colleges of Gauteng Province in South Africa

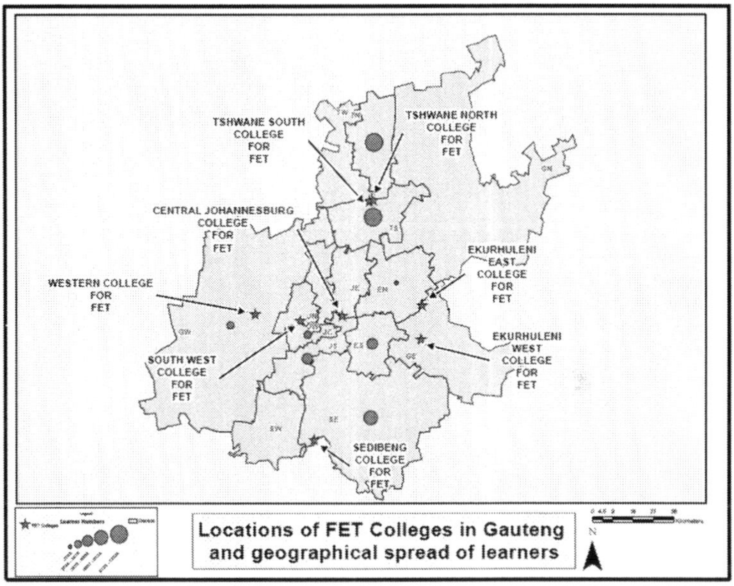

In this study the researcher's selection of the participants was done purposefully, based upon the participant being a member of the college council under the study. College principals were also used because of their expertise in the FET sector. According to Patton (1990), purposeful sampling means selecting information of rich cases that may yield ample information for studying a phenomenon in depth. The advantage of purposeful sampling lies in the selection of knowledgeable participants who are able to provide the required data for the research. Each participant has a unique set of data in a particular area that is needed to form a holistic understanding of the phenomenon. In this research the researcher selected participants who possessed the necessary information and who were willing to share the information with the researcher in a manner that ensured that all areas of information required by the research topic were covered.

McMillan and Schumacher (2001) argued that the power and logic of purposeful sampling, is that a few cases studied in depth may yield many insights about the topic. Patton (1990) supported this view by asserting that qualitative research did not normally make use of large samples. According to him the validity of a sample is not dependent on the number of participants, but rather on the degree of trustworthiness of the participants and their information. Patton (1990) further stated that validity, meaningfulness, and insight generated from the inquiry have more to do with the richness of information of the cases selected and the observational and analytical capabilities of the researcher, than with the sample sizes.

The research used College Councils of the Gauteng Public Further Education and Training Colleges as the subjects for research. As indicated above, there are eight (8) Public Further Education and Training Colleges (FET) in the Gauteng Province. Each FET College is governed by a college council which is made up of sixteen (16) members. The grand total of all council members of the eight colleges are one hundred and twenty-eight (128) members (see Table 1).

Table 1: Number of Respondents per FET College Council

Gauteng FET colleges	No of council members	Council members selected into sample	Percentage
Central Johannesburg College	16	16	12.5
Ekurhuleni East College	16	16	12.5
Ekurhuleni West College	16	16	12.5
Tshwane North College	16	16	12.5
Tshwane South College	16	16	12.5
South West Gauteng College	16	16	12.5
Sedibeng College	16	16	12.5
Western College	16	16	12.5
Total	128	128	100.00

Data collection

Data collection was done by means of a structured, self-completion questionnaire, where respondents were required to select their responses from alternatives provided. The researcher also used face to face interviews to collect data from college principals. The sample size of study was large enough for conducting statistical tests based on the chi-square random variable the Pearson chi-square test of association and binary logistic regression analysis. The researcher also considered that quantitative method is a positivist method for testing the research existing models by deduction and reach a generalisation for the topic. The researcher used qualitative method and as its name implies, described a set of non statistical inquiries and processes used to gather data generally about social phenomena of FET colleges (McNabb, 2002).

Qualitative approach used

The qualitative method used was the in-depth interviews carried by the 8 Principals of the colleges (Barbour, 2008). Objectives and defined questions were the two main factors which decided the research strategy used for this study. The survey used an intensive one-on-one in-depth interview adapted from William (2006). The researcher used survey research as one of the most important areas of measurement in applied social research. The broad area of this survey research encompassed measurement procedures that involved asking questions to respondents. The survey strategy involved administration of questions to a group of 8 principals of the 8 colleges. The researcher further developed a list of higher order questions which were used to principals on one-on-one interviews with the researcher being the interviewer.

Quantitative approach used

Survey research provides the greatest breadth of coverage because of the ability to send surveys to many individuals at many institutions. Surveys require the researcher to be very specific about concepts and measures. Moreover, surveys and subsequent quantitative analysis were perceived to be the most objective method because the researcher was not present during data collection, which reinforced anonymity. This was supported by Boyd (2003) when they wrote that surveys provided data on a large number of people in their varied and normal life settings.

Data analysis

Statistical methods of data analysis adopted both qualitative and quantitative data analysis. Frequency tables and cross-tabs were used for summarising data. Binary logistic regression analysis was performed for various factors that affected corporate sustainability as supported by Hosmer and Lemeshow (2000). The reliability of the fitted logistic regression model was assessed based on standard classification tables, and the Hosmer-Lemeshow goodness-of-fit test. Adjusted odds ratios were used as a measure of effect. Odd ratios were adjusted for potential confounding variables which were gender, age category and race, place of residence and level of education. In-depth interviews were analysed based on tallying, coding and text analysis (Qualitative research). Triangulation with subject matter specialists was used for verification.

Use of ordinal scales for measurement

Ordinal scales were used for measuring subjective attributes such as the quality of services provided to stakeholders and the degree of adherence to recommended guidelines. 5-point ordinal scales were used for measurement in which 1 denoted the lowest score and 5 denoted the highest score. The Cronbach-Alpha test for internal consistency was used to ascertain the reliability of the 5-point ordinal scales (Cronbach-Alpha coefficient = 89%).

5. Results of the Study

Frequency table for categorical variables of study (Table 2)

Y: Corporate sustainability (Yes, No) is the dependent or outcome variable of studyThe dependent variable of study (Y) is dichotomous because its 2 possible values are 1 and 0

Y=1 if factors needed for the efficient and effective corporate sustainability are not available

Y=0 if factors needed for the efficient and effective corporate sustainability are available

Table 2: Frequency Table for Corporate Sustainability Factors

Corporate sustainability	Frequency	Percentage
Yes (Y=0)	23	17.84
No (Y=1)	105	82.16
Total	128	100

Figure 4: Sustainability Factors

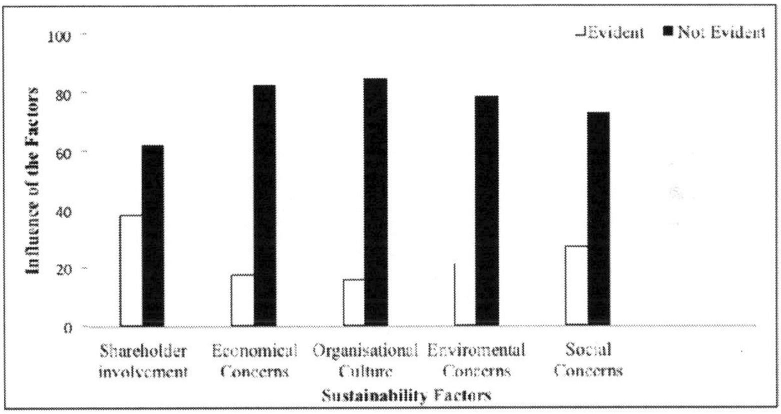

Figure 4 shows that 72% believe that there is no evidence of college boards incorporating in their governance systems and structures much issues of social concerns. College council boards through the respondents reflect that only 21.36% provide evidence of existence of environmental concerns. 78.64% of the council members professed ignorance on dealing with this issue as part of their governance. The figure also reflects that only 27.18 of the respondents showed that there was evident of social concerns in terms of the triple bottom line reporting principle. Such evident further showed that 72.82% of council boards do not even consider integrated reporting in their annual reports. The figure further shows lake of accountability systems and structures as evidenced by a 20.39% consciousness in almost all colleges. The figure also indicates that 84.47% of the council boards are not providing support of the cultural bottom line. 15.53% of the respondents indicate that something to do with the working conditions is being done while the holistic approach to organisation's cultural preservation is being ignored.

Finally, the figure above clearly shows also that there is only 37.86 evidence of the college boards' involvement in colleges' existence. 62.14% shows that there is no framework which permits performance enhancing mechanisms for stakeholder participation.

Cross-tab analysis

The Pearson chi-square test of association was used for identifying significant two-by-two associations among pairs of categorical variables. At the 5% level, significant associations have P-values that are smaller than 5%. All expected cell frequencies were 5 or above. This indicates that P-values obtained from cross tab analysis were reliable. Table 3 shows a list of significant associations which were obtained from Pearson's chi-square tests of associations testing the factors affecting corporate sustainability.

Table 3: Results obtained from the Pearson Chi-square of Associations

Variable of study associated with factors affecting corporate governance sustainability	**P-Value**
Evidence of stakeholder involvement	0.000
Social, concerns being addressed	0.000
Environmental concerns being addressed	0.000
Organizational cultural concerns being addressed	0.000
Economic issues being addressed	0.000

Table 3 model shows that viability of corporate sustainability is influenced by no evidence of stakeholder involvement, Social, concerns not being addressed, Environmental concerns not being addressed, Cultural concerns not addressed, Economic issues not being addressed, in a decreasing order of strength.

Results obtained from binary logistic regression analysis

Adjustment was done for five well known potential confounding variables in the FET college sector of Gauteng Province. These confounding variables were level of education, age category, race, gender and the status of council membership (internal or external). Adjusted and unadjusted odds ratios were similar in magnitude and significance. This shows that none of the five potential confounding variables was a confounding variable. Thus, the estimated results were not affected by confounding variables.

Table 4: Odds ratios obtained from Logistic Regression Analysis Model

Factors that affect sustainability	Odds Ratio	P-Value	(95% Conf. Interval)
No evidence of stakeholder involvement	9.04	0.005	[1.69, 45.42]
Social, concerns not being addressed	15.51	0.005	[2.15, 42.45]
Environmental concerns not being addressed	8.99	0.001	[1.11, 22.59]
Cultural concerns not addressed	11.58	0.002	[1.11; 25.60]
Economic issues not being addressed	12.83	0.003	[1.10; 89.71]

At the 5% level, significant odds rations have P-values that are smaller than 0.05

Table 4 model shows that the odds ratio of the variable "Environmental concerns not being addressed is 8.99." This shows that an FET college that does not address environmental concerns of the area in which it operates is 8.99 times as likely to lack corporate sustainability in comparison with another college that addresses environmental concerns of the area in which the college operates.

Assessment of goodness-of-fit

For Table 4 model, the percentage of overall correct classification was equal to 85.11%. Percentage sensitivity was equal to 89.665 and percentage specificity was equal to 77.78%.

The Hosmer-Lemeshow goodness-of-fit test for Model

Table 5 shows that the P-value obtained from the Hosmer-Lemeshow goodness-of-fit test was equal to 0.3209 > 0.05, thus showing that the model is reliable at the 5% level.

Table 5: The Hosmer-Lemeshow Goodness-of-Fit Test for Model

Number of observations	**128**
Number of covariate patterns	29
Pearson chi2 (22)	22.42
Prob > chi2	0.3209

Results obtained from in-depth interviews (Qualitative)

In-depth interviews showed that the following factors were adversely affecting FET colleges in Gauteng Province: No college had a sustainability policy and strategy or even sustainability plans as a guiding policy to accountability. In order to achieve success, ownership of sustainability must extend beyond senior management team and become a way of thinking throughout the organisation this has not been evident in the interviews carried out with the College Principals. Each college council in Gauteng province agreed that it is their responsibility to execute and develop sustainability goals and plans as part of the annual business planning process although that was not evident.

6. Discussion

College council members and college principals understood quite well that to be successful, all college stakeholders needed to be involved in achieving the colleges' sustainability goals and setting new targets. However, there were no measured steps in college progression to ensure that the pace was right for college council members and management. There was common ground also such stakeholders would bring fresh ideas and perspective to college sustainability journey and colleges needed to look for opportunities to work with them. The research made the following findings:

- The research revealed that any FET college operating within a community is now considered a corporate citizen of that community (Mervyn, 2009).
- The research further revealed that for FET colleges to embrace corporate citizenship, it has to realise all the sustainability factors which are stakeholder involvement, social concerns, environmental concerns, cultural concerns, and economic concerns.

- FET colleges do not have sustainability strategies to support management control, performance measurement, and reward systems for sustainable development.
- FET college councils did not have systems and structures to support corporate sustainability such as policies, committee to report on sustainability issues, include college sustainability as part of the council's annual report. However, sustainability was actually replaced by corporate social responsibility which involved donating food and clothing to the poor in the community.
- Most of the college principals agreed that there was need to embrace corporate citizenship as part of college governance. However, they highlighted the following as some of the issues which needed to be addressed.
    - The composition of the councils was cause for concern as their experience in corporate sustainability was very limited.
    - The only qualified staff members in any board who is the company secretary does not form part of the board hence the board chairperson relied on the principals who also did not have much expertise in administering boards.
    - While the government emphasised on the appointment of chartered accountants in FET colleges, it was imperative for the government again to recommend appointment of company secretaries in the colleges to advise the boards
- The government had given the college councils extended power to develop their own statute within which they would operate as a board for each college. A provisional general statute was provided by the government to just help the colleges while they were developing their own. The research results revealed that none of the colleges had developed a statute in line with the needs of the college. The reasons given during the interviews with principals revealed that there was lack of capacity on what issues needed to be included in the statutes hence they were using the general one provided by the government which was skeletal.

7. Recommendations and Conclusion

The five factors from the sustainable model were critical and influential in determining public colleges' sustainability. Now that the essentiality and comprehensiveness of the sustainability exposure component has been validated, colleges need now turn to the issue of corporate citizenship to further their understanding of how colleges might manage their corporate sustainability. As found in the models of the data analysis, factors found at the extremes of the corporate social performance were also considered very important for college sustainability.

Colleges embracing corporate citizenship would have an institutional oriented top-management philosophy, collaborative external affairs strategy and well-developed internal organisational arrangements. Conversely, colleges with minimal exposure to such would not exhibit enterprise-oriented top-management philosophy, adversarial external affairs strategy and little or no internal organisation systems and structures of sustainability. Colleges embracing corporate citizenship would best be served by a corporate philosophy that is highly integrated with society and long term in its outlook, with an integration of the college's economic interests with its social goals which is a characteristic of strong corporate sustainability.

The researcher therefore makes the following recommendations to the FET colleges in the province of Gauteng and to the Department of Higher Education and Training:

- The college council boards should act as the focal point for corporate sustainability. The boards should identify and take account of the legitimate expectations of stakeholders. Stakeholders should be engaged in such a manner that they have trust and confidence in these public institutions.
- The college boards should be supported by a qualified company secretary if the councils are to appreciate and embrace corporate sustainability as part of good sustainable governance
- College councils should in the interim procure external expert services of a qualified company secretary to assist with putting up systems and structures that are current and in line with sustainable development as enshrined in King III report.

- The college council should identify the organisation's internal and external stakeholders and agree on a policy, or policies to determine on how the college should relate to all stakeholders. Any corporate sustainability framework to be adopted by the colleges should recognise that rights of stakeholders as established by law should be active and clear thereby promoting all stakeholders in creating wealth, jobs, and sustainable public institution.
- A new framework and governance model for FET colleges which consider expertise and experience when appointing council members should also be developed. This would help to develop council self performance mechanisms for the entire council, its committees and individual offices such as that of the chairperson.
- College council boards should ensure that the colleges act as responsible corporate citizens. They (colleges) should not undermine the sustainability of the social and natural environment. As part of corporate citizenship colleges should be sustainable and should enable future generations to meet their needs. All colleges should adopt integrated reporting and should incorporate corporate sustainability factors in their annual reports
- Good corporate citizenship should be integrated into the culture of all colleges. A corporate citizenship framework should be developed by each college to promote corporate sustainability issues in the areas they are situated which should be complemented by proper models of monitoring and evaluation. These should also form part of the government legislation on college governance.
- Sustainability reporting should be focused on substance over form and should transparently disclose information that is material, relevant, accessible, understandable and comparable with past performance of the college and the performance of others.
- The board should ensure that the college develops and implements an effective compliance framework and process.

It is clear that development of a model for evaluating and managing the corporate sustainability of colleges will provide a systematic and valuable direction for college councils and principals interested in discovering how colleges might best manage their corporate sustainability issues and pursue a strategy of corporate sustainability. With respect to the research questions proposed, the research undertaken has been both quantitative and qualitative but it was not sufficient to satisfactorily answer the research questions. So it may not be possible to articulate that factors of corporate sustainability found or lack of in colleges had a greater impact to their performance. What it is possible to state though is that a college which has a more complete understanding of corporate sustainability factors will address issues of corporate sustainability more professionally. By implication a more complete understanding of the inter-relationships will lead to better sustainable governance thereby authentically answering the research questions.

Finally, future research may also want to continue the efforts begun in this study and expand on the results discovered herein. While the expanded dataset found can be used by the government to come up with framework for measuring corporate sustainability, more detailed data is needed to explore within education sector relationships since this research has looked at the main factors only.

References

Adams, J., Khan, H. T., Raeside, R. and White, D. (2007), *Research methods for graduate business and social science student,* London: SAGE.

Argyris, C. (1982), *Reasoning, Learning, and Action,* Jossey-Bass: San Francisco, CA.

Barbour, R. (2008), *Introducing Qualitative Research: a Student's Guide to the Craft of Doing Qualitative Research,* London: SAGE.

Barlett, P. and Chase, G. (2004), *Sustainability on Campus: Stories and Strategies for Change,* Cambridge: Cambridge University Press.

Boyd, D.P. (2003), 'Chicanery in the corporate culture: worldcom or world con?', in *Journal of Corporate Governance,* 3(1):83-85.

Crowther, D. (2002), *A Social Critique of Corporate Reporting,* Ashgate: Aldershot

Edwards, A. (2005), *The Sustainability revolution; portrait of a paradigm shift,* Canada: New Society Publishers.

Elkington, J. (2001), 'The Triple-bottom Line for 21Century Business', *In* Starkey, R. and Welford, R.(eds.), *The Earthscan Reader in Business & Sustainable Development*, London: Earthscan Publications Ltd.

Freeman, E. (1984), *Strategic Management: a Stakeholder Approach,* Boston: Pittman Publishing Company.

Global Reporting Initiative, (2002), *Sustainability Reporting Guidelines,* Boston: GRI.

Gray R., Owen D. and Maunders, K. (1992), *Corporate Social Reporting: Accounting and Accountability*, Prentice-Hall: London.

Hawkin, P. (2007), *Blessed Unres*t, New York: Penguin.

Herremans, I.M., Akathaparn, P. and Mcinnes, M. (1992), 'An investigation of corporate social responsibility, reputation and economic performance', in *Accounting, Organizations & Society,* 18(8):211-233.

Hosmer, D.W. and Lemeshow, S. (2000), *Applied Logistic Regression,* 2nd edition, New York: John Wiley and Sons.

Mervyn. K. E. (2009*), King Report on Corporate Governance for South Africa,* Johannesburg: Institute of Directors in Southern Africa.

McMillan, H. J. and Schumacher, S. (2001), *Research in education: A conceptual introduction,* New York: Longman.

McNabb, D. E. (2002), *Research Methods in Public Administration and non-Profit Management: Quantitative and Qualitative Approach,* New York: M.E. Sharpe.

Patton, M.Q. (1990), *Qualitative evaluation and research methods,* New Delhi: SAGE Publications.

Saiia, D. (2007), *Stakeholder salience, issues management and mapping new ways to sustainability*, London: Routledge Falmer.

Tinker, T. (1988), 'Panglossian accounting theories: the science of apologising in style Accounting', in *Accounting, Organizations and Society*, 13 (2), 165-89.

United Nations Conference on Environment and Development (1992), 'Rio Declaration on Environment and Development', Available at http://www.un.org/documents/ga/conf151/aconf15126-1annex1.htm

Waddock, S. and Graves, S. (2002), 'Responsibility: The new business imperative', in *Journal of Academy of Management Executive,* 16(2): 132-147.

WCED (1987), 'Report of the World Commission on environment and Development', Available at www.un-document.net/wced-ocf.htm

Welford, R. and Jones, D. (1998), 'Beyond environmentalism and towards the sustainable organization', In Welford, R. (ed.), *Corporate Environmental Management: Systems and Strategies*, London: Earthscan Publications Ltd.

Werback, A. (2009), 'When sustainability means more than green', in *Journal of McKinsey Quarterly,* 2009(4): 74-79.

William, M.K. (2006), *Research Methods Knowledge Base,* Cambridge: Cambridge University Press.

CHAPTER 11

Human Capital Development, Technological Capabilities and National Development: The Nigerian Experience

ഗ്ന്ദ

Victor E. Dike

1. Introduction

This paper examines the forces that have prevented the leaders from utilizing effectively the abundant human and natural resources at their disposal to build up the technological base of the society and transform the nation into an industrialized country. The population of Nigeria is estimated to be about154 million and rapidly growing without commensurate economic growth rate and development to sustain the teeming population (NBS, 2009). Unemployment rate is over 14 percent and public schools lack adequate funding to provide quality education to students; and the few private schools in the society are beyond the reach of the common people (NBS, 2009; Kakwagh and Ikwuba, 2010).

And teachers at all levels of the educational system (primary, secondary, and tertiary) are not properly motivated to effectively perform their duties. As a result, they are always on strike to force the government to pay their basic salaries and benefits; the government has over the years been promising to improve their conditions of services, but has each time failed to honour its promises. And the teachers' frequent industrial actions have more often than not disrupted regular academic calendars (NBS, 2009). Consequently, experienced and high quality teachers who are disgusted and cannot any longer tolerate the poor conditions of services they found themselves in and are leaving in droves to other neighbouring countries with better working conditions. These are among the major causes of the dwindling standard of education in Nigeria. Poor investment in human capital development (education and health) and particularly the neglect of technical and vocational education and training (TVET) and science-based technology education has contributed in no small measures a lack of highly-skilled technical manpower and "technological capabilities" (Mahon, 2003) to drive the economy and transform Nigeria into an industrialized society and improve the living conditions of the teeming population.

Experts in human capital development ranging from classical scholars such as Schumpeter (1942) and others, to the contemporary scholars such as Schultz (1993) and Hanushek and Kym, (2005) have identified some dimensions of human capital: entrepreneurship skills (or human skills or human skills capital); technological progress that results from Joseph Schumpeter's 'creative destructive'; and accumulated human knowledge and skills that are as important as physical capital in the production process; and of course social capital (relationship between individual-social or institutional capital. These are some of the positive results of long-term investment in human capital development (education and health) (Piazza-Georgi, 2002).

But poor leadership and governance are the integral part of the causes of the problems facing the nation, as they have made investment in human capital development difficult. Economic growth and development are difficult if not impossible in any society that does not seriously invest in human capital development. And that is the crux of the matter with oil-rich Nigeria where the majority of the political leaders are not working for the well-being of the citizenry but for their selfish purposes. They are known to have diverted public funds allocated to education and other national development programs to their private bank accounts. In addition to the poor investment in human capital development is the neglect of the infrastructure and institutions that drive the economy. Today, these structures are in shambles, leading to unfriendly business environment. Many of the business organizations, particularly manufacturing industries are closing down, leading to rising unemployment and crime. The industries that can afford it are relocating to neighbouring countries with alluring business environment.

It is unquestionably a Herculean task to govern a society, but it becomes much more daunting in a society such as Nigeria with an uncertain sociopolitical and economic environment. The state and fate of Nigeria is today defined by the ineffectiveness of the political leaders, non-functional institutions and dilapidated infrastructure that have crippled the economy and pauperized the people. Thus, the myriad sociopolitical and economic problems facing the society today are exerting untold pressure on the people who are utilizing every available opportunity to pour out their anger and frustration. The question is can the greedy and 'me first' new breed politicians lead Nigeria to true democracy, true capitalistic system, and economic prosperity?

2. Purpose

The purpose of this paper is to examine the obstacles to Nigeria's quest for national development, from the point of view of investment in human capital development, technological capability, and good governance and leadership. Thus, the paper seeks to discuss the issues in an attempt to suggest some solutions to the policymakers.

3. Research Methods

The data for this case study were derived from the research and analysis of scholars, analysts and practitioners, government documents and recent newspaper and journal articles. This is to say that the primary method of study was an extensive review of available literature for an in-depth analysis of the problems facing the Nigerian educational system and the economy. The sources of information were carefully evaluated and analyzed to determine their veracity. The educational system of any society is designed to transmit the beliefs, values, attitudes, and norms from one generation to another as well as to train skilled technical manpower to drive and sustain the economy.

4. Problem Statement

Over the years Nigeria has been investing less on education, infrastructure, and institutions that are critical to building to a viable economy. Yet the political leaders think that Nigeria can be transformed into an industrialized society without these necessary preconditions. The political leaders of Nigeria do not appear to be working for the people's welfare, but for their own. Thus, the people should check their enthusiasm whenever they make their usual paper promises. The developed countries around the globe could not have become what they are today without investing in human capital development, good governance and leadership, and technological capability.

5. Research Questions

This paper focuses on the following research questions: Can Nigeria become an industrialized society without functional fundamental institutions and infrastructure (such as effective judicial system, stable electricity supply, functional educational system, good network of roads, security, good leadership and good governance)? Are Nigerians (leaders and followers) making good efforts to resolve the challenges facing the nation and her quest for national development? Can Nigeria meet her challenges for national development without copious investments in human capital development and technological capabilities? Can the people (the leaders and followers) build the technological base that will transform Nigeria into an industrialized society?

6. Challenges Facing Education and the Nigerian Economy

This paper will discuss the myriad forces that have prevented Nigeria from investing in education and the emerging instructional technologies so as to empower the citizens with employability skills through quality education directed by highly trained and motivated teachers. From the foreground it has convincingly been argued by human capital development gurus that without investment in human capital, physical capital, and social capital, economic growth and development will be difficult if not impossible.

7. Human Capital Development

Both ancient and contemporary human and social capital thinkers have been confronted by the issue of human capital development for decades because they recognize their critical role human capital and physical capital development play in individual worker's productive capability, the prosperity of a nation, and the overall living condition of the people (Schumpeter, 1942; Schultz, 1993; Becker, 1993; Hanushek and Kym, 2005; Bell and Pavitt, 1995).

However, Becker (1993) who is widely known for his work on human capital development observed that human capital theory tends to draw a distinction between general education and specific training. He observed that the key focus of human capital theory is how education increases human productivity by improving human knowledge and skills and increases an individual worker's the productive capacity, future income, and lifetime earnings. Thus, the neglect of this important element has a far-reaching consequence on the welfare of a nation. According to Becker (1993) general education creates general human capital, and technical and vocational education provides specific human capital. In human capital framework the economic prosperity and progress of a nation depends precariously on the stock of its physical and human capital.

This author anchors his work on this perspective because no nation can talk about the manpower planning, growth and development, and the creation of employment to improve the lives of the citizens without bringing education (formal, non-formal and informal) or learning into the equation. This is because the shortage of manpower (this author emphasizes technical education) has the tendency to impede individual worker productivity, national productivity, and economic growth. In other words, human resource development has an essential role to play in the economic health of a nation (Perkins *et al.*, 2001).

Recognizing the importance of education in national development, the UNESCO international experts in their meeting in Bonn, declared that "since education is considered the key to effective development strategies, technical and vocational education and training must be the master key that can..., improve the quality of life for all and help achieve sustainable development" (UNESCO, 2004: 107). One of the major problems facing the Nigerian democracy and economy is that the nation, which is aspiring to transform into an advanced economy, is not investing in the necessary institutional and infrastructural conditions.

Social thinkers in the field of human capital and social capital theory have observed that worker's productivity capability is determined by the many variables, including its human skills and ability, health, motivation, and job satisfaction (Judge, May 2001). Experts in this field as well as social analysts have also observed that the majority of the sociopolitical and economic issues facing Nigeria can be attributed to its poor investment in human capital and physical capital development, poor governance, and visionless leadership (Dike, April-September, 2010; Ferriss, 2006; Lewis, 2004).

Nigeria's underdevelopment status can also be traced to inept leadership, which breeds corruption that has become pandemic in the society (see *San Francisco Chronicle,* 2007; Dike, April-September, 2010). And this appears the undisputable reason for the lack of investment in institutional and infrastructural structures, human capital development (particularly health, technical and vocational education and training, and science-based technology education) that will enable that nation to develop a good stock of highly-skilled technical manpower that drive economic growth and national development and improve the health of the nation (Krueger and Mikael, 2001).

Thus, a lack of long-term investments in human capital and the accumulation of knowledge and human skills and "physical capital development" or the "stock of equipment and structures that are used to produce goods and services" (Mankiw, 2008) constitute a bottleneck to economic growth and development and creation of employment for the citizens (Hanushek and Kym 2005; Becker, 1993). Specifically, poor investment in human capital development and physical capital development (poor electricity distribution, poor potable water supply, poor network of roads, railways, airways, congested sea ports, poor postal system, etc.), has contributed in no small measures to the poor performance of the economy, and by extension, the dwindling standard of education in Nigeria.

Over the years employers of labour in Nigeria have been complaining that most of the recent university graduates lack employability skills and problem-solving abilities and other competencies that are the main variables that determine an individual worker's productivity capabilities (Mankiw, 2008; Dike, 2002). Yet Nigeria's educational administrators and policymakers have done very little, if anything, to improve the quality of human capital, which has direct bearing on workers' productivity (Perkins *et al.*, 2001).

However, because of the role of 'human skills capital' (HSK) on individual productivity and by extension their profit margins, organizations around the globe are said to be spending billions of dollars annually to educate and upgrade their employees' 'human skills capital' (Piazza-Georgi, 2002). More often than not workers are rewarded for their higher productivity and the high quality of their goods and services, which influences their firms' profits margins (Clayton, 1995). A journey back to the memory lane shows that before independence in 1960 the schools in Nigeria were properly funded and managed and the students were well behaved and the quality of graduates was equal to those from advanced nations (Dike, 2002). But today, the schools lack adequate funding to hire quality teachers' and purchase the necessary instructional technologies to produce high quality graduates and enhance technological development. Studies show that no nation can make any meaningful socioeconomic and political stride without a good pool of skilled technical manpower (Hanushek and Kym, 2005) and technological capabilities (Mohan, 2003; Bell and Pavitt, 1995).

Instead of Nigeria's schools to be properly funded and developed into a place for high quality research, technological innovation, and for good character formation, they have become a breeding ground for cultism and other destructive social behaviours. Today, 'values education', good 'moral values' and civic responsibility, which shape the character of nations, have become a thing of the past (Dike, 2002). As Lickona (1992) has aptly observed 'respect and responsibility are the two foundational moral values' that a society should teach its citizens. Others admirable qualities include honesty, fairness, tolerance, prudence, self-discipline, helpfulness, compassion, cooperation, courage (the virtues of Aristotle) among other democratic values. Thus, 'taking responsibility for the things we do wrong as well as the things we do right' (Lickona, 1992) is important for national development.

As noted earlier, as the nation's quality of education is rapidly deteriorating so is also the economic environment. In its 2010 report on 'Doing Business in Nigeria 2010-Through Difficult Times', the World Bank ranked Nigeria 125 out of 182 economies surveyed in its Global Doing Business Report. The report showed that about 90 per cent of Nigerian businesses that operate in the informal sector lack access to credit. One would add that similar problems are being faced by those in the formal sector of the economy. And in its 2008 Review of World Development, the United Nations Development Programme (UNDP) ranked Nigeria 157 out of 177 in Human Development Index; Nigeria was also among the 'Least Liveable' nations (UNDP, 2008).

Some of the problems facing Nigeria today require common sense solutions, but others would involve sophisticated framework of ideas and application of advanced technologies. As Albert Einstein observed decades ago, "the specific problems we face today cannot be solved at the same level of thinking we were at when we created them" (see quote from Albert Einstein, n/d). Thus, for the economy to grow and for businesses to thrive and create employment, Nigeria should employ or elect political leaders with proper skills, knowledge and vision as well as political courage to adopt and implement policies to meet the challenges facing the nation.

8. Leadership and Governance

The issue of the role leadership in the direction of organizations and development of nations has continued to prompt debates as leadership appears to have many meanings (Marquis and Huston, 2012). Although there is no single definition of leadership it is imperative for this paper to consider some perspectives. Ward (2009) has defined leadership as "the art of motivating a group of people to act towards achieving a common goal" and not a selfish objective. And as Northouse (2007: 3) has aptly noted, leadership "is a process of getting things done through people"; it "means responsibility" - having "passion for the purpose and the mission of the organization" or society one leads.

However, because of poor leadership or 'leadership without a purpose' the state of the polity appears to be rapidly deteriorating because very few of the leaders of Nigeria, if any, work to improve the living conditions of the citizenry. Over the years a wave reform programs have been undertaken, but the society still lacks the leadership with political will and commitment to implement effective policies to transform the economy and tackle the sociopolitical problems facing the nation. The political leaders do not seem to understand that leadership is assuming responsibility for something. They appear good at prescribing solutions to economic problems (Dike, 2003) without providing the institutional and infrastructural framework to help the economy grow (Acemoglu, 2003; Edison, 2003).

And more often than not, their policies are hastily put together and poorly executed. The political landscape is littered with wreckage of unreasoned policies and those involved in such activities appear to enjoy the nation's underdeveloped status (Dike, 2003). Given the foreground, the activities of the leaders shape the reality of the nations they lead (UNDP, 2008). To resolve the leadership challenge facing Nigeria today and transform its organizations into productive entities (Kouzes and Posner, 1995) Nigerians must elect or appoint men and women of proven integrity, drive and know-how to transform (Burns, 2003) and manage the affairs of the nation. This is because without meeting the ethical and leadership challenges (Johnson, 2005) and improving organizational effectiveness (Bass and Avolio, 1994) Nigeria may not achieve her objectives. The nation's economic growth and development rests with good leadership and governance. Thus, for Nigeria to move forward the leaders as well as the followers must clean up their act and invest in the fundamental infrastructure and institution that will promote technological innovation, transform the economy, and develop the society.

Related to the leadership problem is governance, which has

among others, been defined as a system of values, policies, and institutions by which a society manages its economic, social, and political affairs through interactions within the state, civil society and private sector. Put differently, governance comprises the mechanisms and processes for citizens and groups to articulate their interests, to work together and mediate their differences, and exercise their legal rights and obligations with rules, institutions and practices that set limits and provide incentives for individuals, organizations and firms (UNDP, 2008). Thus, good governance (political, economic and social governance) - the three dimensions of governance (Shabbir-Cheema, May 2005) refers to the question of how a society can organize itself to ensure equality of opportunity and equity (social and economic justice) for all citizens. Good governance promotes people-centered development; but 'bad governance', which is the opposite of good governance (Shabbir-Cheema, May 2005) is among the major cause of the problems facing Nigeria. For instance, the people are not allowed equal economic opportunity and freedom to participate in the political process. And this is threatening to destabilize the polity and undermine Nigeria's democratization process.

According to Sen (1999), 'unfreedoms' leave the people with little choice to exercise 'their reasoned agency, and "Freedoms are not only the primary ends of development, they are also among its primary means". Thus, development whether social, political and economic, "requires the removal of major sources of unfreedoms" (Sen, 1999: 36-37). Because of bad governance (and thus corruption) Nigeria's system lacks checks and balances (or mechanisms) to control the autocratic tendencies in government and to hold political actors accountable for their actions. The politicians do not practice ethical politics and their actions do not add values to the system. Lack of 'ethical politics and values' and politics of hate and destruction have contributed significantly to the economic and political hiccups in the society. Corruption (Smith, 2008) is, however, a greater part of the problems facing Nigeria as it leads to bad governance that has hampered socio-political and economic development.

9. Bribery and Corruption

Although corruption is a global scourge, Nigeria appears to suffer the most from it because the leaders are pathologically corrupt. Is Nigeria born to be corrupt? Everyone appears to believe that Nigeria has a 'culture of corruption' (Smith, 2008). Over the years, Nigeria has earned a lot of revenue from crude oil, which has gone down the sinkhole created by corruption. An article in the *San Francisco Chronicle*, 11 March 2007, described Nigeria as a rich nation floating on oil wealth ,but almost none of it flows to the people' "Oil giant that runs on grease of politics." This is evidence in the fact that top public servants are very rich because they harbour the mentality that public money belongs to no one (Dike, April-September, 2010; October-March, 2011).

Nigeria appears a hotbed of corruption and the so-called 'war against corruption' seems like trying to contain the wind. The country has become synonymous with corruption. Majority of the elected officials, appointees, and top government are corrupt. They often extract bribes from the citizens (business owners, those seeking to do business with the state, etc.) while performing their official duties. The magnitude of high-profile political corruption in Nigeria tends to dwarf any genuine efforts to enhance economic growth and nation development and diminishes the peoples' trust in government. National dailies are awash with news of how public officials are acquiring million-dollar homes (within and outside Nigeria) and stockpiling stolen public money in financial institutions abroad. Thus, corruption scandals are strewn all over the society like a straw hut in a hurricane (*BusinessDay,* 14 May 2010; 15 May 2010; and 17 April 2012).

There are two sides to every problem; private individuals are also involved in corrupt activities. A spate of studies shows that corruption leads to "poor governance and low growth" and hampers social development (Rose-Ackerman, 2004). And the leaders of Nigeria have demonstrated their unwillingness to reform the society despite the number of top government officials involved in corruption (*BusinessDay,* 14 May 2010; 15 May 2010; and 17 April 2012). For instance, the former governor of Delta State, James Ibori, who eluded the Economic and Financial Crimes Commission (EFCC), was arrested in Dubai recently (*BusinessDay,* 14 May 2010) on money laundering charges.

Since corruption is an anathema to political stability and economic prosperity this author can hazard to conclude that Nigeria's problems are not from the "harshness and the niggardliness of nature" (Keynes, May 1932), but corruption that prevented the society from investing in human capital development, particularly technological capabilities that would drive the economy, create employment, and put the nation's abundant human and resources into productive use.

If the political leaders want to fight corruption they should start by fixing the economy because some of the petty corrupt activities are caused by poverty. For instance, the roads and highways in the society are roadblocks manned by police officers who are virtually thieves in uniform; they are ill-trained and poorly paid (Dike, April-September, 2010; October-March, 2011). In addition, to breathe a new life into the nation's ossified economy Nigeria must strengthen her critical institutions and infrastructure that drive the economy.

10. Infrastructural and Institutional Challenges

Any person familiar with Nigeria would agree that among the challenges facing the economy is ineffective institutions and dilapidated infrastructure (bad roads, erratic power supply, limited access to potable water and basic healthcare, and ineffective regulatory agencies, et cetera). The plethora of reforms and development policies are ineffective due to shoddy infrastructure and institutional failure (Hoff and Fall, 2003). In particular, electricity supply is poor and expensive; and investors are not impressed. In a country where electricity takes about 40 percent cost of production improving electricity supply would lure foreign investors into the society and spur the sluggish economy (*Daily Sun,* 07 June 2010). Without functional infrastructure and effective institutions and with good education system Nigeria can be turned into the Singapore of West Africa. Singapore is said to have 'one of the world's best-educated populations and busiest port' (*The Economist*, 25 February 2012: 80).

As economists have observed, building a vibrant economy or restoring growth to an ailing economy takes resources. To ensure long-term growth and prosperity, Nigeria must use its resources wisely, invest in science-based technology education to train a good stock of highly-skilled technical manpower to build and maintain the critical institutions and infrastructure without which the economy will not gain from the "power of productivity" (Lewis, 2004). A nation enjoys higher standards of living if the workers can produce large quantities of quality goods and services for local consumption and extra for export (Mankiw, 2008). Without technological capabilities, however, the economic will remain in shambles, productivity will remain low, the quality of goods and services will remain poor, and Nigeria will be unable to produce enough food to feed its teeming population and will not effectively compete in the global marketplace.

11. Technological Capabilities

This paper posits that the problems facing the Nigerian economy (including the low capability utilization of domestic industries and firms) hinge precariously on the nation's poor investment in human capital development (education and health) and a lack of long-term investment in education, particularly science-based technology education and technical and vocational education and training (TVET) that will enable the country to develop a good stock of highly-skilled technical manpower training to provoke system innovation, technological development and economic growth that will sustain its industrialization process.

Put differently, the 'growth tragedy' (Easterly and Levin, April 1997), of the Nigerian economy has its roots in ineffective and inefficient domestic development policy design and implementations and the failure of the policymakers to learn from the impressive history of economic growth and development of the East Asian countries, that are today branded Asian Tigers. Thus, to move forward technologically Nigeria has to copiously invest in technical education and develop "an effective national system of innovation" (Mohan, 2003), which is a precondition for rapid technological development. The pace of technological change in any society is a reflection of the health of its infrastructure and institutions, and of courses its social capital, which will encourage innovative system activities and enhance industrial capital accumulation and technological transformation and national development (Mohan, 2003).

As in human capital development, physical capital development in the form of technological capabilities and innovation will drive a sagging economy and create employment. Some studies show that there is a correlation between the underdeveloped status and soaring unemployment in developing nations, including Nigeria, to their lack of technological capabilities and a good stock of highly-skilled technical manpower to drive the economy and create employment (Rose, 2009; Mohan, 2003; Sen, 1975; Dike, October-March, 2011).

Thus, because of a lack of investment in science-based technology education and technical and vocational education and training (TVET)Nigeria is suffering from a shortage of highly-skilled technical manpower to build and maintain its critical infrastructure and produce quality goods and services at a least cost and compete effectively in the global marketplace. And the on-going global financial and economic crisis that started in 2008 has worsened the already bad economic situation in Nigeria, including rising youth unemployment (Kakwagh and Ikwuba, 2010), social crisis and lawlessness (e.g. kidnapping for ransom and the *Boko Haram*, a militant Islamic group) onslaught (*Daily Trust*, 18 August 2010; *ThisDay,* 09 January 2012).

Social scientists, especially economists have aptly noted that modern economy demands technological skills to enhance innovation and to function effectively and efficiently (Rose, 2009; Offor, 2007; Freeman, 1987). In particular, development economists have noted that socio-economic changes or transformations are impossible in any society without the preconditions, one of which is technological capabilities (Webb and Grant, 2003; Soubbotina, 2004). In addition, studies show that no nation has developed without 'technological capabilities' (Kim, 1980, 1997; Bell and Pavitt, 1995; Mohan, 2003).

For Kim (1980) "technological capability" is "the ability to make effective use of technological knowledge in efforts to assimilate, use, adapt and change existing technologies" (Kim, 1980: 254-277). In fact, experts have noted that 'technological capability' is an essential component for "sustainable development", which is a 'development' that 'meets the needs of the present generation without compromising the ability of future generations to meet their own needs' (United Nations, 11 December 1987; Smith and Rees, 1998). In spite of some flickers of development efforts (some people would argue that the efforts are not meaningful-that they are mere political rhetoric), the economic environment remains very unfriendly with industries and organizations still facing a huge hurdle. However, because of the poor employment opportunities most of the youth are frustrated and desperate to leave the country. Stories abound of how desperate Nigerians, mostly youth, take extraordinary risks to travel to Europe, the United States and even to some African nations in search of an elusive better life. Some of them have been arrested, tortured, and maimed or imprisoned in foreign countries; and some of them have died in the deserts of North Africa in the process of trying to escape economic hardship (Hamoon, 13 May 2006; Fletcher, *BBC News*, 05 December 2007).

12. Concluding Remarks

System Innovation and Technological Capabilities

Does Nigeria possess the needed skilled technical manpower for system innovation and technological transformation? The world economy is rapidly changing and the only countries that are capable to compete effectively in the global marketplace are those that are investing in human capital development and technological innovation. Nigerians are tired of being hungry in the midst of plenty. Investors (both domestic and foreign) are not motivated to invest in the domestic economy because of the unlimited hurdles posed by the shoddy infrastructure and institutions.

It is impossible for Nigeria to meet her technological and system challenges and move along with the changing global economy without investing copiously in technological education and research and development. To transform her economy and effectively compete in thc global marketplace tomorrow Nigeria should begin today to nurture and strengthen the infrastructure and institutions that makes technological development possible. Technological innovation will give rise to new entrepreneurs and knowledge-driven industries that will push down the barriers to economic growth and development.

Thus, no nation can grow and thrive in the present knowledge-intensive global economy without a long-term investment in human capital development (education and health), system and technological innovation. No society can grow and thrive in the present twenty century knowledge-driven global economy without copious investment in human capital development (education and health), research and development (R&D) and technological innovation either by developing serious indigenous technological innovation (fresh thinking) or copying and adapting other nation's technological ideas in the cheapest possible means.

China and other serious nations have gone through this route to become technological giants. And today, China possesses the technological clout to gain from the power of productivity (*The Economist,* May 26th – June 1st, 2012). To move forward, Nigeria should adopt and adapt the economic policies and technological innovation of successful nations. Nigeria can become an industrialized nation either by developing her indigenous technology or copy and adapt other nation's technological ideas by the cheapest possible means.

Thus, without creating a hospitable environment to lure investors into the economy; and without ensuring that the Constitution and rule of law is respected and business contracts honoured; and without investing in human capital development, particularly technology education Nigeria will continue to dream of being an industrialized nation.

References

Acemoglu, D. (2003), 'Root Causes: A historical approach to assessing the role of institutions in economic development', in *Finance and Development*, June, 40 (2): 27-30.

Bass, B.M. and Avolio, B.J. (1994), *Improving Organizational Effectiveness through Transformational Leadership*, Thousand Oaks, CA: Sage.

Becker, G.S. (1993 [1964]), *Human Capital: A Theoretical and Empirical Analysis, with Special Reference to Education*, 3rd edition, Chicago: University of Chicago Press.

Bell, M. and Pavitt, K. (1995), 'The Development of Technological Capabilities', In Hague, I. U. (ed.), *Trade, Technology and International Competitiveness*, Washington, DC: The World Bank.

Burns, J.M. (2003), *Transforming leadership,* New York: NY: Grove/Atlantic, Inc.

Business Day (2010), 'EFCC seeks Ibori's extradition from Dubai', May 14.

*Business Day (*2010), 'I resigned to face corruption charges-Ogbulafor', May 15.

Business Day (2012), 'Ibori jailed for 13 years', April 17.

Clayton G.E. (1995), *Economics: Principles and Practices*, New York: Glenco, McGraw-Hill.

Daily Trust (2010), 'Experts Proffer Solution to Collapse of Buildings', August 18.

Daily Sun (2010), 'In Nigeria, power is about 40% cost of production – John Aluya, MD, TechnoGlass', June 7.

Dike, V. E. (2002), 'The State of Education in Nigeria and the Health of the Nation', in *NESG Economic Indicators,* Jan-March, 8(1): 45-50

Dike, V. E. (2003), 'Nigeria: Economic Growth and Institutional Factors', in *NESG Economic Indicators,* October–December, 9 (4): 46-50.

Dike, V.E. (2010), 'Review of the Challenges facing the Nigerian Economy: What is the Way Forward?' in *NESG Economic & Policy Review,* 17(2&3): 45-53.

Dike, V.E. (2011), 'Can Nigeria build its economic future?', A Critical Review of Issues Facing the Nation', in *Economic and Policy Review,* 17&18(4&1): 24-32.

Edison, H. (2003), 'Test the Links: How strong is the links between institutional quality and economic performance?' in *Finance and Development*, June, 40(2): 35-37.

Easterly, W. and Levin, R. (April 1997), 'Africa's Growth Tragedy: Policies and Ethic Divisions', Available at:

http://williameasterly.files.wordpress.com/2010/08/17_easterly_levine_africasgrowthtragedy_prp.pdfon (Accessed: 17 February 2012).

Einstein, A. (n/d), Available at: http://www.quotedb.com/quotes/11 (Accessed: 28 March 2011).

Ferriss, A.L. (2006), 'A Theory of Social Structure and the Quality of Life', in *Applied Research in Quality of Life,* May, 1(1): 117-123.

Fletcher, P. (2007), 'Mirage of rich Europe Lures Job-Hungry Africa', *BBC News-online*, December 5.

Freeman, C. (1987), *Technology and Economic Performance: Lessons from Japan,* London: Pinter.

Hamoon, S. (2006), *Africa Transit Migration through Libya to Europe: the Human Cost,* May 13, Cairo: FMRS/AUC.

Hanushek, E. and Kym, D. (2005), 'Schooling Labour Forces Quality and Economic Growth', *NBER Working Paper 5399*, Cambridge, MA.

Hoff, K. (2003), 'Paths of Institutional Development: A view from Economic History', in *The World Bank Research Observer,* Fall, 18(2): 205-226.

Judge T.A., Thoresen, C.J., Bono, J.E. and Patton, G.K. (May 2001), 'The job satisfaction–job Performance relationship: A qualitative and quantitative review', in *Psychological Bulletin,* 127(3): 376-407.

Johnson, C.E. (2005), *Meeting Ethical Challenges of Leadership*, 2nd edition, Thousand Oaks, VA: Sage

Kakwagh, V.V. and Ikwuba, A. (2010), 'Unemployment in Nigeria: Causes and Related Issues', in *Canadian Social Science,* 6(4): 61-67.

Keynes, J.M. (1932), 'The World's Economic Outlook.', *The Atlantic,* Available at: http://www.theatlantic.com/unbound/flashbks/budget/keynesf.htm (Accessed: 04 December 2006).

Kim, L. (1980), 'Stages of Development in Industrial technology in a Developing Country', in *Research Policy*, 9(3): 254-277.

Kim, L. (1997), *From Imitation to Innovation: The dynamics of Korea's Technological Learning*, Boston, MA: Harvard Business School Press.

Kouzes, J.M. and Posner, B.Z. (1995), *The Leadership Challenge: How to keep getting extraordinary things done in Organizations*, San Francisco: Jose-Bass.

Krueger, A.B. and Mikael, L. (2001), 'Education for Growth, Why and for Whom?' in *Journal of Economic Literature,* 39(4): 1101–136.

Lewis, W.W. (2004), *The Power of Productivity: Wealth, Poverty, and the Threat to Global Stability*, Chicago: University of Chicago Press.

Lickona, T. (1992, October), *Educating for Character: How Our Schools Can Teach Respect and Responsibility,* New York: Bantman Books.

Mankiw, N. G. (2008), *Essentials of Economics,* 5th edition, South-Western: CENGAGE Learning: Australia.

Marquis, B.L. and Huston, C.J. (2012), *Leadership Roles and Management Functions in Nursing.* New Delhi: Wolters Kluwer, Lippincott Williams and Wilkins.

Mohan R. (2003), *Facets of the Indian Economy: The NCAER Golden Jubilee Lectures,* New Delhi, India: Oxford University Press.

National Bureau of Statistics (NBS) (2009), 'Federal Government of Nigeria-Statistical Fact Sheet on Economic and Social Development', Nigeria: Abuja.

Northouse, G. (2007), *Leadership theory and practice,* 3rd edition, Thousand Oak, London, New Delhi: Sage Publications.

Offor, E.N. (2007), 'Refocusing Science and Technology Education toward National Development', in *Knowledge Review: A Multi disciplinary Journal of NAFAK,* 15 (6): 53-55

Perkins, D.H., Radelet, S., Snowgrass, D.R., Gillis, M., and Roemer, M. (2001), *Economic Development,* 5th edition, W.W. Norton Company. See in particular Chapters 8 and 9.

Piazza-Georgi, B. (2002), 'The role of human and social capital in growth: extending our Understanding', in *Cambridge Journal of Economics,* 26: 461-479.

Rose-Ackerman, S. (2004), 'Governance and Corruption', In Lomborg, B. (ed.), *Global Crises, Global Solutions*. Cambridge: Cambridge University Press.

Rose, K. (2009), 'Enhancing Innovation in Developing Country Systems: A Synthesis of Case Studies and Lesson from Uganda', in *African Journal of Science and Technology, Innovation and Development,* 1(2&3): 89-114.

San Francisco Chronicle (2007), 'Oil giant that runs on grease of politics', March 11.

Schumpeter, J.A. (1942), *Capitalism, Socialism and Democracy*: Harper (reprinted by Harper Colophon, 1975).

Schultz, T.W. (1993), *Origins of Increasing Returns,* Oxford, Blackwell.

Sen, A. (1975), *Employment, Technology and Development*, Oxford: Clarendon.

Sen, A. (1999), *Development as Freedom*, New York: Anchor Books (see Chapters 1-3), pp.36-37.

Shabbir Cheema, G. (2004), 'From public administration to governance: the paradigm shift in the link between government and citizens', *Paper presented at the 6th Global Forum on Reinventing Government: Towards Participatory and Transparent Governance*, 24-27 May 2005.

Smith, C. and Rees, G. (1998), *Economic Development*, 2nd edition, Basingstoke: Macmillan.

Smith, D.J. (2008), *A Culture of Corruption,* Princeton and Oxford: Princeton University Press.

Soubbotina, T.P. (2004), *Beyond Economic Growth: An Introduction to Sustainable Development,* 2nd edition, Washington DC.: World Bank Publications.

The Economist (2012), 'Business in Rwanda- "Africa's Singapore?"' p. 80.

The Economist (May 26 - June 1, 2012), 'How strong is China's economy?' p. 11.

This Day (2012), 'Jonathan-Boko Haram Has Sympathizers in Govt', January 9.

United Nations (1987), 'Report of the World Commission on Environment and Development', General Assembly Resolution 42/187, December 11. Available at: http://www.un-documents.net/a42r187.htm (Accessed: 06 September 2010).

UNESCO (2004), *Learning for Work, Citizenship and Sustainability,* Final report, UNESCO International Experts Meeting, Bonn, Germany, 25-28 October 2004:107.

UNDP (2008), 2008 Review of World Population, New York: The United Nations.

Ward, S. (2009), 'Leadership', About.com. Canada: Small Business; Available at: http://sbinfocanada.about.com/od/leadership/g/leadership.htmon (Accessed: 18 March 2012.

Webb, C. J. and Grant, C. (2003), 'Economic Transformation: An Implementation Framework', *Tertiary Education Commission Research Paper*, Wellington: Tertiary Education Commission.

CHAPTER 12

Post Scriptum:
Human Capital to Shape African Future

Mammo Muchie & Angathevar Baskaran

It is in 1961 that the term human capital appeared when the nobel laureate economist Theodore W. Schultz wrote the article "Investment in Human Capital." There is a need to invest in and to build human capital to be able to acquire and deploy knowledge, talent, skills and lessons from employing and using without interruption human resources to find tangibe ways to remove unemployment, poverty and inequality.

In Africa it appears that it is mainly the African youth and women who appear to live a costly life. It is said that 60% of the jobless are between 15-24 years old in the areas of Africa South of the Sahara. 72% of the youth are said to live on less than $2 a day and 46% in less than a $1 a day. It is also reported that about 90% of Africa's jobs are in the informal economy. Those who work in the informal sector cannot have high productivity and high quality and skilled jobs. The level of poverty amongst the youth and women are reported to remain high despite the fact that we have now the Africa rise narrative. There is now a positive account that many African countries GDPs are growing at over 5% annually. Science, Technology, Engineering and Mathematics (STEM) education to create technical skills to make the youth employable is critically needed to remove the high poverty levels. We do not only have unemployed youth but unemployable youth making building human capital very necessary to transform the lives of the African youth along with all those engaged in work too.

The Africa union has created a youth and student council to make young people creators, inventors, innovators and entrepreneurs. There has been also an Africa entreprenurship award to stimulate business start ups in all the countries of Africa. Over 2000 start-ups have been able to produce business plans for mentors to evaluate the applications. A number of them were selected for the awards. This award stimulates a Pan-African entrepreneurship value and culture to influence the youth to come out of poverty and to become job creators rather than job seekers.

Undoubtedly, it is human capital that is the means for the transformation of the youth to acquire innovative, creative and entrepreneurial culture. When we teach engineering post graduates in Africa, we incentivise them to learn how to create productivity from given in-puts by increasing the output. Lack of technical and innovative skills has to be removed. To empower and to entitle the youth, an enabling environment that fosters the entrepreneurial acumen and the inven-tive spark have to be synergised to create high quality jobs to make the youth the assets rather than the liability for African economies.

The skills mismatch creates job seekers that do not find jobs. There are limiltations in the relationships between industry and university. There are graduates that do not find jobs, and there are industries that seek graduate employees. This is due to the skills mismatch proving that human capital is not managed in a way that motivates, encourages and stimulates skills match.

The African youth is expected to reach one billion by 2050 and Africa has put forward an African agenda to realise full unity by 2063! At the moment Africa has the lowest school enrolment in the world. Nearly 100 million young people roam about searching for jobs anywhere. Some have employment, but it is as good as not be-ing employed since it is in the informal sector with low skill- jobs.

There is a need to make a paradigm shift from low skills to high skills, from quantity to quality and from lack of human capital to abundant human capital for promoting sustainable African development. Access to quality education for all the youth must be considered a human right and all children and teenagers must acquire quality education that can earn them employment. They must not finsh university education and come out and they have no jobs. It must be the opposite. They finish school and they have jobs either by their own creation of jobs or by being employed without any doubt. The youth suffering from the curse of unemployment, poverty, inequality and underemployment can only lead to crime and conflict. When the youth enjoys quality education human capital becomes the foundation for producing high quality goods and services.

As the world is entering the 4th industrial revolution, Africa cannot continue to remain in the global value chain producing and trading in raw materials and agricultural production. The human capital for raw materials and agriculture is low skilled and Africa needs to move to the knowledge economy and the digital age. Africa has to upgrade the value chain through knowledge and skills and remove as much as possible poverty, unemployment and inequality.

One of the benefits of the 4th industrial revolution is that many young people have gone digital. Africa is said to be able to have mobile phone coverage for all by 2025 making communication, transferring finance and other services easier across Africa. It is like having a digital infrastructure to facilitate productive interactions across Africa. There should be a parallel double movement where human capital grows as the population of Africa grows as well. If the population grows it can be either an asset or a liability or risk depending on whether the young people acquire quality education to make a positive difference to their lives. Human capital is critical to make the Afican youth movers and change makers. The talent of the African youth needs to be harnessed to make a big difference to Africa's future. Human capital management is the high way for making the African population the shakers and movers of the African future and destiny. Let all the stakeholders realise and act to make a big difference to the lives of the people and nature.

Notes on Contributors

Mammo Muchie

DST/NRF SARChI Professor (Innovation & Development), Tshwane University of Technology, South Africa; Technology and Management Centre for Development (TMCD), Oxford University, UK; Associate Faculty Professor, Sussex University, UK; University of Technology (UoG), Arsi University & Adama Science and Technology University (ASTU), Ethiopia. Email: MuchieM@tut.ac.za

Angathevar Baskaran

Associate Professor, Department of Development Studies, Faculty of Economics and Administration, University of Malaya, Malaysia; Senior Research Associate of SARChI (Innovation & Development), Tshwane University of Technology, Pretoria, South Africa. Email: baskaran@um.edu.my & anga_bas@yahoo.co.uk

Victor E. Dike

Adjunct Professor, School of Engineering and Technology, National University (Sacramento Center), and Business/Information Technology Instructor at the Fremont School for Adults, (Sacramento City Unified School District) Sacramento, California. Email: vdike@cwnet.com

Joseph M. Kizza

The University of Tennessee-Chattanooga, 615 McCallie, Chattanooga, Tennessee, 37403, USA. Email: joseph-kizza@utc.edu

Timothy Waema

University of Nairobi, Kenya, P.O. Box 30197, 00100, Nairobi, Kenya. Email: waema@uonbi.ac.ke

Myron N.V. Williams

Center for Functional Nanoscale Materials and Department of Chemistry, Clark Atlanta University, Atlanta, GA 30314, USA. Emails: Myron N.V. Williams - mnwill@cau.edu

Paul M. Brown

Office of Study Abroad, Clark Atlanta University, Atlanta, GA 30314, USA. Email: Paul M. Brown – pbrown@cau.edu

Ishrat M. Khan

Center for Functional Nanoscale Materials and Department of Chemistry, Clark Atlanta University, Atlanta, GA 30314, USA. Emails: Ishrat M Khan - ikhan@cau.edu

Alfred Msezane

Center for Functional Nanoscale Materials and Department of Physics, Clark Atlanta University, Atlanta, GA 30314, USA.
Email: Alfred Msezane – amsezane@cau.edu

James Reed

Center for Functional Nanoscale Materials and Department of Chemistry, Clark Atlanta University, Atlanta, GA 30314, USA. Emails: James Reed - jreed@cau.edu

Laurisa London

Center for Functional Nanoscale Materials and Department of Chemistry, Clark Atlanta University, Atlanta, GA 30314, USA. Emails: Laurisa London- laurisalondon@gmail.com

Malik Maaza

Center for Functional Nanoscale Materials, Clark Atlanta University, Atlanta, GA 30314, USA; iThemba LABS, West Somerset, South Africa. Email: Malik Maaza - maaza@tlabs.ac.za

Nwaocha Vivian Ogochukwu

School of Science and Technology, National Open University of Nigeria, 14/16 Ahmadu Bello Way, Victoria Island, Lagos State, Nigeria. Email: webdevee@yahoo.com

Thabo I Pudi

Campus Director, Tshwane University of Technology, eMalahleni Campus, South Africa. Emails: puditi@tut.ac.za and puditi@vodamail.co.za

Vuyisile Msila

Faculty of Education and Nursing, University of Johannesburg, South Africa. Email:vmsila@uj.ac.za

Cheryl le Roux

Department of Further Teacher Education, University of South Africa (UNISA). Email:lrouxcs@unisa.ac.za

Agnieta B Pretorius

Department of Information and Computer Technology, Tshwane University of Technology, Email:PretoriusAB1@tut.ac.za

Coneth Richards

Faculty of Engineering, Tshwane University of Technology, Email:RichardsCG@tut.ac.za

Marius Dieperink

Faculty of Humanities, Tshwane University of Technology, Email: DieperinkM@tut.ac.za

Ahmed Driouchi

Institute of Economic Analysis & Prospective Studies (IEAPS), Al Akhawayn University, Ifrane, Morocco. Email: A.Driouchi@aui.ma

Nada Zouag

Institute of Economic Analysis & Prospective Studies (IEAPS), Al Akhawayn University, Ifrane, Morocco. Email: N.Zouag@aui.ma

Boniface Ngah Epo

Lecturer, Department of Economics and Management, University of Yaoundé II, Cameroon. P.O. Box 1365 Yaoundé. E-mail: epongahb@yahoo.fr

Francis Menjo Baye

Associate Professor, Department of Economics and Management, University of Yaoundé II, Cameroon. P.O. Box 1365 Yaoundé. E-mail: bayemenjo@yahoo.com

Manager M. Muswaba

Social Scientist, WorldFish Center, P.O. Box 51289, Ridgeway, Lusaka, Zambia, Email: e.madzudzo@cgiar.org or emadzudzo@gmail.com.

Zeleke Worku

Professor, Business School, Faculty of Management Sciences, Tshwane University of Technology, Pretoria, South Africa, 159 Skinner Street, Pretoria 0001, South Africa, Email: workuz@tut.ac.za

Index

L

M

N

Q

R

U